STEREO COMPUTER GRAPHICS

and Other True 3D Technologies

PRINCETON SERIES IN
COMPUTER SCIENCE   *David R. Hanson and Robert E. Tarjan, Editors*

The Implementation of the Icon Programming Language by Ralph E. Griswold and Madge T. Griswold (1986)

Recent Advances in Global Optimization edited by Christodoulos A. Floudas and Panos M. Pardalos (1992)

The Implementation of Prolog by Patrice Boizumault (1993)

Stereo Computer Graphics and Other True 3D Technologies edited by David F. McAllister (1993)

STEREO COMPUTER GRAPHICS
and Other True 3D Technologies

Edited by David F. McAllister

PRINCETON UNIVERSITY PRESS · PRINCETON, NEW JERSEY

Copyright © 1993 by Princeton University Press

Published by Princeton University Press, 41 William Street,
Princeton, New Jersey 08540
In the United Kingdom: Princeton University Press, Chichester, West Sussex

Library of Congress Cataloging-in-Publication Data

Stereo computer graphics and other true 3D technologies / edited by
David F. McAllister.
p.   cm.
Includes bibliographical references.
ISBN 0-691-08741-5
1. Computer graphics.   2. Three-dimensional display systems.
I. McAllister, David F., 1941–
T385.S758   1993
006.6—dc20   93-16642

This book has been composed in Times Roman by Publication Services

Princeton University Press books are printed on acid-free paper and meet the guidelines for
permanence and durability of the Committee on Production Guidelines for Book Longevity
of the Council on Library Resources

Printed in the United States of America

10 9 8 7 6 5 4 3 2 1

To my son, Timothy Walt McAllister

Contents

Preface

This work grew from notes produced by the contributors for several half- and full-day courses in true 3D technologies and techniques. These courses have been presented at both SPIE and SIGGRAPH for several years. The book is the product of the efforts of many individuals, but my student and colleague Larry Hodges, now at Georgia Tech, was a prime mover behind this effort and deserves much of the credit for the book's inception.

Our motivation is to share our knowledge and experience with researchers and practitioners who wish to compare 3D techniques and technologies or to implement 3D algorithms and interfaces in computer graphics applications. We assume that the reader is familiar with basic freshman-level physics and "three-dimensional" computer graphics, including the topics of affine transformations, projections, clipping, hidden-surface elimination, and illumination.

The term *3D* in computer graphics has different meanings for different audiences. A common meaning is the projection of a three-dimensional scene onto a flat surface, where both the left and right eyes see the same image; there is no *binocular disparity*. Many authors use the term *2.5D* to describe this case and to distinguish it from the true 3D case. We will use the term *3D* to mean an image in which the left- and right-eye views are slightly different and in which (most) viewers perceive depth.

I wish to thank Steve Wixon, James Lipscomb, and Michael Starks for their helpful advice; Vicki Fels, Mark Waldrop, Cherie Bucklew, and Rhonda Covington for their help in producing the final manuscript; Lou Harrison for his aid in editing the final manuscript; Dorothy Strickland and Nancy Long for their help in editing page proofs; and my contributors for their willingness to produce rewrite after rewrite.

The Contributors

Philip Bos, currently a principal scientist at Tektronix, received his Ph.D. in the area of liquid-crystal physics from Kent State University in 1978. Following a year of postdoctoral work at Kent's Liquid Crystal Institute, he joined Tektronix in 1979. At Tektronix he has investigated nematic and ferroelectric liquid-crystal devices, and he invented a rapid-switching liquid-crystal shutter device called the π-cell. He has also been involved in applications of liquid-crystal devices, particularly in the areas of field-sequential color displays and field-sequential stereoscopic displays.

Jessie Eichenlaub founded Dimension Technologies in 1986 and serves as vice president of research and development. He has responsibility for scientific and technical activities including research, new product specifications, and patent development. He has been involved in stereo technology since 1977 and has five patents to his credit, with several applications pending. He holds a B.S. in physics from the Rensselaer Polytechnic Institute and has taken graduate courses in lens design from the University of Rochester.

Lou Harrison received his B.S. in computer science from North Carolina State University in 1987 and his M.S. in computer science, also from NCSU, in 1990. Mr. Harrison has taught courses in operating systems and computer graphics at NCSU and is currently employed as software systems manager for the Department of Computer Science at NCSU as he pursues his Ph.D. He has done research in surface generation for computer-aided milling and autostereoscopic display technology. He is currently exploring lossy compression techniques applied to stereo images. Harrison is a member of ACM, SIGGRAPH, IEEE, and SPIE.

Larry F. Hodges is an assistant professor in the College of Computing at Georgia Institute of Technology. He received his Ph.D. in computer engineering at North Carolina State University in 1988. He also holds a M.S. in computer science/engineering from NCSU, a M.A. in religion from Lancaster Theological Seminary, and a B.A. with a double major in mathematics and physics from Elon College. His research and consulting interests are in computer graphics, stereoscopic display, virtual environments, and scientific visualization. He is a member of ACM, IEEE-CS, SID, and SPIE.

Edwin R. Jones is a professor in the Department of Physics and Astronomy of the University of South Carolina, Columbia. He is the author of an electronics textbook and an introductory college physics text. His research interests include holography and other forms of three-dimensional imaging. His work in three-dimensional imaging has led to four U.S. patents. He served for three years on the university's patent committee as its chairman. Jones received his

B.S. degree from Clemson University and his M.S. and Ph.D. degrees in physics from the University of Wisconsin, Madison. He is a member of the American Physical Society, the American Association of Physics Teachers, and SPIE.

Lenny Lipton studied physics at Cornell University. He has made documentary films, he has written articles and columns for national magazines, and he wrote the popular song "Puff, the Magic Dragon." His book *Independent Filmmaking,* published by Simon & Schuster in 1972, is the standard text on the subject. *Foundations of the Stereoscopic Cinema* was published by Van Nostrand in 1982.

He founded StereoGraphics Corporation in 1980 and is the company's chief technical officer and chairman of the board. He is the inventor of the flickerless multiplexing technique, which is the basis for the modern electronic stereoscopic display, and is the principal inventor of the CrystalEyes system. He has been granted ten patents and is a member of SMPTE, SID, SPIE, and ASCAP.

Shaun Love received his bachelor's degree from the University of North Carolina at Chapel Hill and his master's and Ph.D. in computer science from North Carolina State University. He is currently doing research in the printer division of Lexmark International.

The editor, **David F. McAllister,** is professor of computer science at North Carolina State University in Raleigh. He received his B.S. in mathematics from the University of North Carolina at Chapel Hill in 1963, his M.S. in mathematics from Purdue University in 1967, and his Ph.D. in computer science from UNC–Chapel Hill in 1972. He has worked in computer graphics since 1970. His interest in 3D technologies began in 1985 with a grant to study the state of the art of 3D for the Defense Mapping Agency. He has presented several tutorials for SPIE and SIGGRAPH in 3D technologies. He is a member of ACM, SIGGRAPH, IEEE, Eurographics, SPIE, and SID.

Porter McLaurin is an associate professor of media arts and former chairman of the Department of Media Arts at the University of South Carolina, Columbia. He earned his B.A. at the University of Florida and his M.Ed. and Ph.D. degrees from the University of South Carolina. He has produced more than one hundred television programs and has been involved in three-dimensional imaging for the past fourteen years. His work in three-dimensional imaging has led to four U.S. patents. He teaches courses in three-dimensional imaging technologies, television production, and communications research. He presently serves on the university's patent committee as its chairman. McLaurin is a member of SPIE.

Lawrence D. Sher graduated from Drexel University with a B.S. in physics and from the University of Pennsylvania with a Ph.D. in biomedical engineering. After eight years on the faculty at Pennsylvania, which included a three-year grant to study ski binding design and a one-year teaching sojourn in Iran, he joined Bolt Beranek and Newman in Cambridge, Massachusetts, in 1971.

His work at BBN has ranged from early medical applications for computers to teaching the use of BBN's software products. His major work has been the conception, creation, and continuing refinement of the vibrating-mirror type of display, christened "SpaceGraph" display in 1976. Embodying an unusual mixture of mechanics, acoustics, optics, and computer systems engineering, the Space-Graph display has been a multidisciplinary, long-term project. Sher has directed this project from concept to commercial reality.

Richard Steenblik is currently immersed in two startup high-technology companies, Chromatek and Applied Physics Research, which focus on innovative consumer-oriented optics. Prior to these ventures he worked for seven years with the Georgia Tech Research Institute as a research engineer on a variety of projects, including solar energy research, energy conservation, the development of automated irregular parcel-post sorting equipment, computer analysis of a military helicopter airframe, and the development of novel low-cost lens fabrication processes.

Steenblik earned his bachelor's degree in mechanical engineering from Georgia Tech in 1980. Since then he has passionately pursued the invention of unusual optical devices, including a digital sundial and a low-cost, high-performance pocket microscope. He holds a total of five U.S. patents, including three covering the chromostereoscopic process, and currently has three patents pending.

Homer B. Tilton's work with 3D CRT displays began in 1949 with the independent conception, design, and construction of an oscilloscope to generate stereo images on a pair of type 3JP1 CRTs. In 1965 a monocular 3D display, the "scenoscope," developed by the author was marketed by Optical Electronics. The scenoscope featured perspective projection and manually controlled scene rotation using all-analog circuitry, as opposed to the mostly digital circuitry used with today's computer-generated monocular 3D displays. In one scenoscope model, the observer was tracked and the resulting position information used to modify the CRT image so that one could partially "look around" the 3D scene simply by moving from side to side. In 1968 the "parallactiscope," a true 3D oscilloscope, became the focus of the author's 3D CRT work. This instrument is described in the text.

Recently (1991) he has begun serious work on a new concept in 3D TV systems, in which a single sensed image and an auxiliary "depth" video signal (one carrying point-to-point depth information) are combined at the receiving end to synthesize the multiple images required for autostereoscopic and autoparallactic true 3D TV displays.

R. Don Williams holds a Ph.D. in industrial engineering, with a specialty in human factors engineering. He has worked for over twelve years in the information display area and has focused in the last five years on the development of alternative technologies for 3D displays. He is the coinventor of Texas Instruments' Omniview 3D displays and has worked as the principal investigator, research scientist, and manager of this new product development effort.

He previously supervised and supported the human factors programs in the military, symbolic processing, office automation, industrial systems, and manufacturing domains. Among his many professional activities he has served as a member of the technical program committee for the Society for Information Display (SID) and coauthored the ANSI standard for the human factors engineering of visual display terminal workstations. Williams has authored twenty-one technical papers and has filed twelve patents.

Yei-Yu Yeh received her Ph.D. in psychology from the University of Illinois at Urbana-Champaign in 1986. Subsequently she was an assistant professor in the Department of Psychology at North Carolina State University. She joined the Display Human Factors Research Group at Honeywell in 1987 and served as a principal scientist until 1990. She is now a faculty member in the psychology department at the University of Wisconsin at Madison. Her research interest focuses on visual coding and attention in three-dimensional space.

STEREO COMPUTER GRAPHICS

and Other True 3D Technologies

1

Introduction

David F. McAllister

Since the publication of Okoshi's famous work [OKOS76], and particularly over the past ten years, there has been rapid advancement in 3D techniques and technologies. Hardware has both improved and become considerably cheaper, making real-time and interactive 3D available to the hobbyist, as well as to the researcher in such areas as molecular modeling, photogrammetry, flight simulation, CAD, visualization of multidimensional data, medical imaging, and stereolithography. The improvements in speed, resolution, and economy in computer graphics, as well as the development of liquid-crystal polarizing shutters and liquid-crystal parallax barrier methods, make interactive stereo an important capability. Old techniques have been improved, and new ones have been developed. True 3D is rapidly becoming an important part of computer graphics, visualization, and virtual-reality systems. In addition, the improvement of high-resolution color printing has made 3D hardcopy more available and useful for archiving and transporting 3D images.

Computer graphics algorithms that work well for producing single monoscopic frames do not necessarily extend easily to rendering consistent, unambiguous stereo pairs. In addition, it is not necessarily the case that the work required to render a left- and right-eye stereo view is twice that required to render a single-eye view. Research in the area of computer-generated stereo animation has only recently begun and has produced many interesting questions. We will treat some of these issues here. Although our emphasis will be computer-generated images, many of the techniques and technologies described also apply to images produced through other techniques, such as photographic methods. Our emphasis is on CRT-produced or reflected images. (See [ROBI91] for a discussion of stereo issues and the optics involved in head-mounted displays.) It is assumed that the reader is familiar with elementary computer graphics at the level of [HILL90] or [FOLE90] and with freshman-level optics.

The following section discusses the common visual depth cues and how they interrelate. Some of these cues will be treated in more detail in other chapters in this book.

1.1 Depth Cues

The human visual system uses many depth cues to disambiguate the relative positions of objects in a 3D scene. These cues are divided into two categories: physiological and psychological. (See [OKOS76], [JULE71], and [GOLD89].)

Physiological depth cues include the following:

Accommodation. *Accommodation* is the change in focal length of the lens of the eye as it focuses on specific regions of a 3D scene. The lens changes thickness due to a change in tension from the ciliary muscle. This depth cue is normally used by the visual system in tandem with convergence.

Convergence. *Convergence,* or simply *vergence,* is the inward rotation of the eyes to converge on objects as they move closer to the observer. Outward rotation of the eyes beyond the normal parallel position for observing objects at a distance is a phenomenon called *wall-eyed* vision. Stereo images that are not correctly registered or computed may force the viewer to look at the image wall-eyed, which can introduce eye strain and subsequent headache.

These two properties will be mentioned numerous times throughout the text. Some technologies produce 3D images in which there is consistency between the two cues, and others do not. We will use these properties to differentiate various technologies in Section 1.2.

Binocular disparity. *Binocular disparity* is the difference in the images projected on the left and right eyes in the viewing of a 3D scene. It is the salient depth cue used by the visual system to produce the sensation of depth, or *stereopsis*. Binocular disparity is modeled in computer graphics with the eyes considered as two off-axis centers of perspective projection. Chapter 5 will discuss these issues in more detail.

Motion parallax. *Motion parallax* provides different views of a scene in response to movement of the scene or the viewer. Consider a cloud of discrete points in space in which all points are the same color and approximately the same size. Because no other depth cues can be used to determine the relative depths of the points, we move our head from side to side to get several different views of the scene (called "looking around"). We determine relative depths by noticing how much two points move relative to each other: as we move our head from left to right or up and down, the points closer to us appear to move more than points further away.

Psychological depth cues include the following:

Linear perspective. *Linear perspective* is the property that the size of the image of an object on the retina changes in inverse proportion to the object's change in distance. This is modeled in computer graphics by a perspective versus parallel projection. Parallel lines moving away from the viewer, like the rails of a train track, converge to a vanishing point. As an object moves further away, its image becomes smaller. This is closely related to the depth cue of retinal image size.

Shading and shadowing. The amount of light from a light source illuminating a surface is inversely proportional to the square of the distance from the light source to the surface. Hence, faces of an object that are further from the light source are darker, which gives cues of both depth and shape. Shadows cast by one object on another also give cues as to relative position and size. Shadowing in computer graphics is very important in producing realistic images.

Aerial perspective. Objects that are further away tend to be less distinct, appearing cloudy or hazy. Blue, having a shorter wavelength, penetrates the atmosphere more easily than other colors. Hence, distant objects sometimes appear bluish. We can simulate the haziness phenomenon in computer graphics through the technique of *depth cueing*.

Interposition. If one object occludes (hides or overlaps) another, we assume that the object doing the hiding is closer. This cue is simulated in computer graphics by the removal of hidden surfaces and hidden lines.

Retinal image size. We use our knowledge of the world, linear perspective, and the relative sizes of objects to determine relative depth. If we view a picture in which an elephant is the same size as a human, we assume that the elephant is further away since we know that elephants are larger than humans.

Texture gradient. We can perceive detail more easily in objects that are closer to us. As objects become more distant, the texture becomes blurred. Texture in brick, stone, or sand, for example, is coarse in the foreground and grows finer as the distance increases.

Color. The fluids in the eye refract different wavelengths at different angles. Hence, objects of the same shape and size and at the same distance from the viewer often appear to be at different depths because of differences in color. (The concept of *chromostereopsis* is discussed in Chapter 4, and a technology based on it is described in Chapter 10.) In addition, bright-colored objects will appear to be closer than dark-colored objects.

The human visual system uses all of these depth cues to determine relative depths in a scene. In general, depth cues are additive; the more cues, the better able the viewer is to determine depth. However, in certain situations some cues are more powerful than others, and this can produce conflicting depth information.

1.2 A Technology Taxonomy

The history of 3D displays is well summarized in several works and hence will not be reviewed here. Okoshi's important work [OKOS76] and that of Tilton [TILT87] each present an excellent history of 3D technologies. Those interested in a history beginning with Euclid will find [MORG82] of interest. In addition, a useful reference that summarizes many patents in the area of 3D technologies can be found in [STAR91]. The contributors include some of the appropriate history in their introductions.

Most 3D displays fit into one or more of three broad categories: *holographic, multiplanar,* and *stereo pair.* In general, holographic and multiplanar images produce "real" or "solid" images, in which binocular disparity, accommodation, and convergence are consistent with the apparent depth in the image. They require no special viewing devices and hence are called *autostereoscopic.*

Autostereoscopic Technologies

Most readers are familiar with holographic displays, which reconstruct solid images. Many of these images have the "lookaround" property; the observer can move his or her head to see a different view of the 3D scene. A popular combination of computer-generated stereo pairs and holography is the holographic stereogram, in which multiple stereo pairs are recorded in strips on a piece of holographic film. An example is shown in Colorplate 1.1.

Another method uses vertical lines or *parallax barriers* to block the left-eye image from the right eye and vice versa. Lenticular displays use vertically aligned lenses called lenticules to direct the appropriate view to each eye. In these two techniques the image is recorded in strips behind the parallax barrier or the lenticular sheet. Although the techniques are old, recent advances in printing and optics have increased their popularity for both hardcopy and autostereoscopic CRT devices. They are described in more detail in Chapter 3. Recently, parallax barrier liquid-crystal imaging devices have been developed that can be driven by a microprocessor and used to view stereo pairs in real time without glasses. The technology is described in Chapter 9. The moving slit is a variation on this theme and is described in Chapter 8.

Multiplanar methods are similar to volumetric methods in computer graphics; the image is subdivided into voxels, or three-dimensional cubes. The varifocal mirror technique divides a 3D scene into thousands of planes, and a point-plotting electrostatic CRT plots a single point from each. A circular mirror vibrating at 30 Hz reflects these points while changing their apparent distance from the viewer, and the points combine to produce a real image. An implementation of a varifocal mirror is described in Chapter 11, and a rotating-mirror display device is described in Chapter 13. In both cases the image is transparent.

Stereo Pairs

The production of stereoscopic photographs (stereo pairs or *stereographs*) began in the early 1850s. (An illustrated history of stereography complete with a price guide can be found in [WALD91].) Stereo pairs simulate the binocular disparity depth cue by projecting distinct flat images to each eye. There are many techniques for viewing stereo pairs; a familiar example is the View-Master reel. Many of the technologies described in this book are based on the stereo pair concept.

Terminology. *Horizontal parallax* is the distance between the left- and right-eye views of a point in a scene projected on a plane perpendicular to the observer's line of sight, called the *stereo window* or *stereo plane*. Positive parallax occurs if the point appears behind the stereo window, the left-eye view being to the left of the right-eye view (see Figure 1.1). Zero parallax occurs if the point is at the same depth as the stereo window, and negative parallax occurs if the point is perceived to lie in front of the stereo window. Normally the stereo window is positioned in the plane of the viewing screen, but it need not be. The perceived depth of a point in a stereo pair depends primarily on the horizontal parallax.

Free-Viewing Stereo Pairs. With practice, most readers can view stereographs without the aid of projecting devices using a technique called *free viewing*. There are two types of free viewing, which are distinguished by how the left- and right-eye images are arranged. The colorplate stereo pairs in this book have been arranged for both *parallel,* or *uncrossed,* and *transverse,* or *cross,* viewing. The leftmost image is a left-eye view, the middle image is a right-eye view, and the rightmost image is again a left-eye view.

For parallel viewing, place a piece of paper between the two leftmost images so that the right eye cannot see the left-eye view and vice versa. Allow your eyes to relax so that the two views merge in the center. The two views in Figure 1.2 are arranged for parallel viewing.

For cross viewing, use the two rightmost images; the right-eye view is on the left, and the left-eye view is on the right. Stare at the region between the two images and gently cross your eyes until the two images merge in the center. The two views in Figure 1.3 are set up for cross viewing. You will still be able to see the right- and left-eye images with your peripheral vision, but concentrate on the center image. Most people will be able to perceive depth in the image after a few seconds.

Figure 1.1
Positive and negative horizontal parallax.

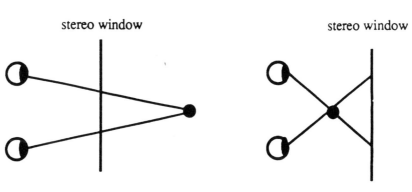

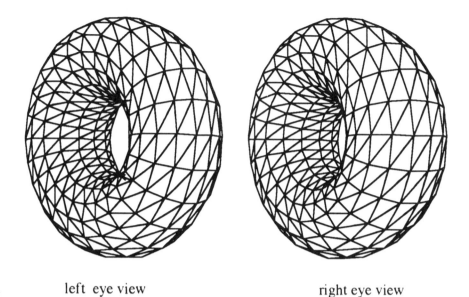

Figure 1.2
Stereo images
arranged for
parallel viewing.

left eye view   right eye view

You must hold the images parallel with your eyes, or you will not be able to merge the left- and right-eye views. Try both techniques on Figures 1.2 and 1.3. If you cross-view Figure 1.2, the depth will be reversed, resulting in *pseudo stereo*. Similarly, if you parallel-view the image in Figure 1.3, the depth will be reversed. Try both techniques on the stereo pair in Colorplate 1.2. Use the free-viewing technique that is most effective for you. Farsighted people find parallel viewing easier, and nearsighted people find

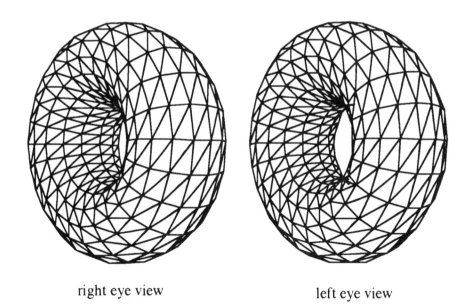

Figure 1.3
Stereo images
arranged for
cross viewing.

right eye view   left eye view

cross viewing easier. If you cannot perceive depth using either technique, you may be a person who is *stereoblind* and cannot *fuse* stereo images (interpret as a 3D image rather than two separate 2D images).

Figure 1.4 is a *random dot autostereogram*, in which the scene is encoded in a single image as opposed to a stereo pair [TYLE90a]. There are no depth cues other than binocular disparity. About 2 percent of the population will not be able to fuse such an image. Using cross viewing, merge the two dots beneath the image to view the functional surface. Crossing your eyes even further will produce other images. (See [BARN91] for a description of how to generate these interesting images.)

Parallax and convergence are the primary vehicles for determining perceived depth in a stereo pair; the observer focuses both eyes on the plane of the display window, which is normally a CRT screen or a projection screen. Hence, accommodation is fixed. In such cases accommodation and convergence are said to be "disconnected," and the image is "virtual" rather than "solid." This inconsistency between accommodation and convergence can make stereo images difficult for some viewers to fuse.

Computation of Stereo Pairs. Several methods have been proposed for computing stereo pairs. Certain perception issues eliminate some techniques from consideration.

Vertical displacement occurs when the left- and right-eye views of a given point, called *homologous* points (Figure 1.5), on a 3D scene differ vertically (i.e., do not both lie on a horizontal line parallel with the viewer's eyes), producing *vertical parallax*. Prolonged viewing of images that have vertical displacement can produce headaches, eye strain, and other uncomfortable

Figure 1.4

A random dot autostereogram of $(-\cos\sqrt{x^2 + y^2})$ on $[-10, 10] \times [-10, 10]$.

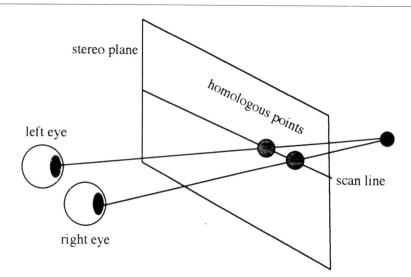

Figure 1.5
Homologous
points.

physical symptoms. A common technique for computing stereo pairs involves the rotation of a 3D scene about an axis parallel to the sides of the viewing screen followed by a perspective projection, which can cause vertical displacement. Hence, this technique is not recommended.

In addition, linear perspective is important for maintaining depth relationships and object shapes. Although parallel projection will not produce vertical displacement, the absence of linear perspective can create a "reverse" perspective as the result of a perceptual phenomenon known as Emmert's law: objects that do not obey linear perspective can appear to get larger as the distance from the observer increases (see Chapters 4 and 7). The preferred method for computing stereo pairs is to use two off-axis centers of perspective projection (corresponding to the positions of the left and right eyes), which simulates the optics of a stereo camera where both lenses are parallel. Chapter 5 discusses these concepts and methods for calculation of stereo pairs in more detail.

1.3 Stereo Output Technologies

Technologies for producing 3D images can be divided into two broad groups: those that present both eye views simultaneously, or *time-parallel* methods, and those that present the two eye images in sequence using optical techniques to occlude the right-eye view when the left-eye view is being presented, and vice versa, called *field-sequential* or *time-multiplexed* methods.

Time-Parallel Techniques

3D movies traditionally used the old anaglyph method, which required the user to wear glasses with red and green filters. Both images were presented on the screen simultaneously; hence, it was a time-parallel method. Many observers suffered headaches when leaving the theater, which gave 3D, and stereo in particular, a bad reputation. A phenomenon called *ghosting* or *cross talk* was a significant problem. The filters did not completely eliminate the opposite-eye view, so that the left eye saw not only its image but sometimes part of the right-eye image as well. This phenomenon is discussed further in Chapter 4. The stereo pairs in this book present both eye views simultaneously and hence use a time-parallel method. This technique can also be used in computer graphics. Hand-held viewing devices permit the adjustment of one eye view to register both the left- and right-eye views so that they can be viewed in parallel. Viewing static images in this way is straightforward, but viewing animation can be difficult and tiring. An early technique for viewing stereo images on a CRT was the half-silvered mirror originally made for viewing microfiche [LIPS89]. The device had polarizing sheets, and the user wore polarized glasses that distributed the correct view to each eye.

Currently the most common way to present stereo pairs to a large audience is by placing orthogonally oriented polarizing filters in front of two projectors. The projection screen must be coated with a metallic surface so that the polarization of the reflected light is maintained. The viewer then uses passive polarized glasses in which the polarization is consistent with the filters on the projectors so that the right-eye image is blocked from the left eye and vice versa. Polarizing filters can also be attached to glass-mounted slides, obviating the need for attaching filters to the projectors. Incorrect positioning of the projectors relative to the screen can cause *keystoning,* in which the image is trapezoidal or V-shaped instead of rectangular. Keystoning causes distortion and produces fusing difficulties for some viewers.

Field-Sequential Techniques

The most popular method for viewing stereo on graphics workstations is the field-sequential technique. The technology for field-sequential presentation has progressed rapidly. Early mechanical devices (now obsolete) were used to occlude the appropriate eye view during CRT refresh. Examples are the rotating disk by Evans and Sutherland and the rotating drum by Bausch & Lomb, originally intended for vector displays because the device was too slow for raster displays. A comparison of many of these devices can be found in [LIPS89].

A technology for workstations uses liquid-crystal shutters that attach to the front of the CRT. The shutter polarizes the left- and right-eye images in different directions, and the user wears passive polarized glasses as described previously and shown in Figure 1.6. Technologies have also been developed in which the polarizing shutters are built into the viewing glasses and powered by batteries. The shutters are switched using an

Figure 1.6
Liquid crystal shutter system (courtesy of Lennie Lipton, StereoGraphics).

infrared signal synchronized to the refresh rate of the CRT. The left- and right-eye views are displayed alternately (at 120 Hz to preclude flicker). This LCD technology is described in Chapter 6.

1.4 Overview

The first half of the book is devoted to stereo techniques. In Chapter 2 Lipton covers perception issues and how they relate to scene composition. He also discusses the concept of orthostereo, which is important in virtual-reality environments where the computer-generated scene must be isomorphic to the real scene being modeled. In Chapter 3 Love covers recent advances in hardcopy methods, including the important parallax barrier technique. In Chapter 4 Yeh treats problems of stereo perception and discusses issues concerning the technologies of Chapter 6. In Chapter 5 Hodges and McAllister compare various algorithms that have been used for computing stereo pairs, including the off-axis projection technique. In Chapter 6 Bos discusses the liquid-crystal technologies that are commonly used on workstations to produce a stereo image. In Chapter 7 Harrison and McAllister treat input devices and software implementation issues for stereo.

The second half of the book discusses other 3D technologies, some of which are stereo-based and most of which are autostereoscopic. Chapter 8 describes a variant of the parallax barrier technique called the moving slit. It was first proposed by Collander and then implemented by Homer Tilton, who contributed the chapter. Chapter 9 is a description of a CRT-based parallax barrier display recently developed by Eichenlaub at Dimension Technologies. In Chapter 10 Steenblik discusses the interesting concept of chromostereopsis, in which a transparent medium such as a prism is used to refract light and create binocular disparity in a colored image. Different wavelengths refract at different angles, causing colors to separate in depth when an image is viewed with special glasses. In Chapter 11 Sher gives a description of the BBN varifocal mirror originally developed by Traub. In Chapter 12 Jones and McLaurin describe the VISIDEP technology, which uses field-sequential vertical parallax to produce a 3D image. Both eyes view the image simultaneously. Williams in Chapter 13 describes the rotating mirror developed at Texas Instruments.

The availability of economical and powerful workstations, and improvements in optics, engineering, printing, and projection techniques have combined to make true 3D more than just a graphics oddity. The technologies and techniques described in this book are becoming a common and important part of the graphics community. We hope the concepts presented here will be useful to both the researcher and the practitioner.

2

Composition for Electrostereoscopic Displays

Lenny Lipton

2.1 Introduction

Until relatively recently, human beings were unaware of the existence of a distinct binocular depth sense, stereopsis. Through the ages, individuals such as Euclid and Leonardo understood that we see different images of the world with each eye. But it was Wheatstone who, in 1838, explained to the world through his stereoscope and drawings that there is a unique depth sense, stereopsis, produced by retinal disparity. Kepler and others wondered why we don't see a double image of the visual world. Wheatstone explained that the problem was actually the solution, demonstrating that the mind fuses the two planar retinal images into one through stereopsis ("solid seeing") [WHEA38].

2.2 Review of Some Depth Cues

Retinal Disparity

When you look at your finger in front of your face, your eyes converge on your finger (Figure 2.1). That is, the optical axes of the eyes cross on the finger. Sets of muscles move your eyes so that an image of the finger is positioned on each fovea, the central portion of the retina. If you continue to converge your eyes on your finger, you'll notice that the background appears to be double. On the other hand, when you look at the background, your eyes converge on it, and your finger, with introspection, will appear to be double (Figure 2.2).

If the images on your left and right retinas could be superimposed like photographic slides, they would overlap almost perfectly except for differences related to your left and right perspective viewpoints. Physiologists call these differences disparity.

Parallax

A stereoscopic display differs from a planar display in only one respect: it provides parallax values of the image points. Parallax produces disparity

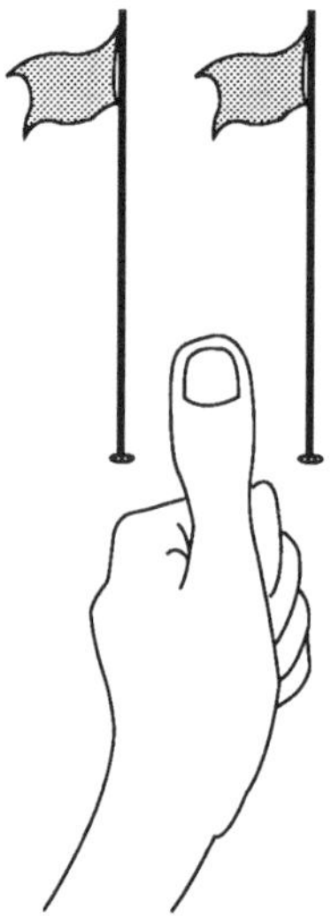

Figure 2.1
Single thumb.
The eyes
converge on the
thumb, and (with
introspection) the
background is
seen as a double
image.

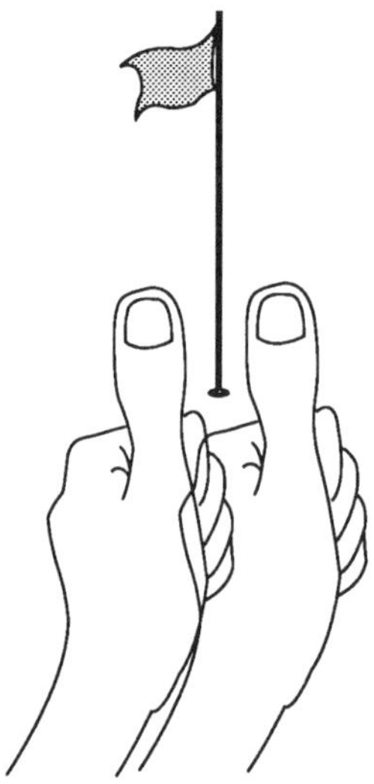

Figure 2.2
Double thumb.
The eyes
converge on the
background,
and (with
introspection) two
thumbs are seen.

in the views presented to the eyes, providing the stereoscopic cue. The stereoscope is a means for presenting disparity information to the eyes, and in the century and a half since Wheatstone it has been improved. The View-Master (Figure 2.3) is one such device. The View-Master stereoscope uses two separate optical systems. Each eye looks though a magnifying lens to view a slide, as in the stereoscope of Brewster (Figure 2.4). The slides are taken with a pair of cameras mounted on a base, replicating the way we see the world with two eyes.

Electrostereoscopic displays provide parallax information to the eye using a method related to that employed in the stereoscope. In the active-glasses display, manufactured by StereoGraphics, Tektronix, and others, or in the on-screen switching device using passive polarizing glasses, manufactured by Tektronix, the left and right images are alternated rapidly on the monitor screen. The distance between left and right homologous, or *conjugate*, image points is the parallax and may be measured in inches or millimeters. Parallax and disparity are similar entities. Parallax is measured at the display screen, and disparity is measured at the retinas.

For a stereoscopic display, when the distance t between the eyes is equal to the parallax, the axes of the eyes will be parallel, just as they are when viewing a distant object. Using parallax values equal or nearly equal to t for a small-screen display will usually produce discomfort.

Any value of parallax between t and zero will produce images that appear to be within the space of the cathode ray tube (CRT), behind the screen. Such objects are said to be within *CRT space*. Objects with negative parallax are said to be within *viewer space*.

Another kind of positive parallax occurs when images are separated by some distance greater than t. This causes wall-eyed vision, a concept men-

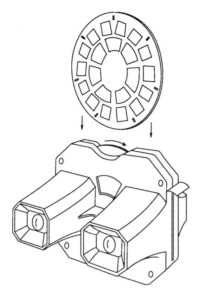

Figure 2.3
View-Master stereoscope.

tioned in Chapter 1, in which the axes of the eyes diverge. This divergence does not occur when you look at objects in the visual world, and the unusual muscular effort needed to fuse such images may cause discomfort. There is no valid reason for divergence in computer-generated stereoscopic images.

Interaxial Separation

The distance between the lenses used to take a stereoscopic photograph is called the *interaxial separation*, with reference to the lenses' axes. The axis

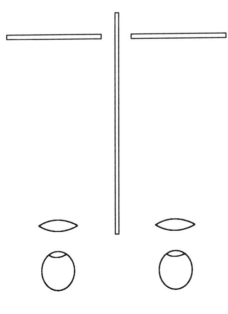

Figure 2.4
Lenticular stereoscope, invented by Sir David Brewster in 1844 and named to distinguish it from the mirror stereoscope. The eyes view the left and right images (stereo pairs) through lenses, which help focus them.

of a lens is the line through the optical center of the lens and perpendicular to the plane of the imaging surface. Whether the discussion concerns an actual lens or the construct of a lens used to create a computer-generated image, the concept is the same. If the lenses are close together, the stereoscopic depth effect is reduced. In the limiting case the two lenses (and axes) correspond, and a planar image results. As the lenses are moved farther apart, parallax increases, and so does the strength of the stereoscopic cue.

2.3 Controlling Depth Cues in Stereo Displays

Congruent Nature of the Fields

Left and right image fields must be congruent in all aspects except horizontal parallax to avoid viewer discomfort [LIPT84b]. The color, geometry, and brightness of the left and right image fields should match within a very tight tolerance, or the result will be eyestrain for the viewer. If a system is producing image fields that are not congruent in these respects, there are problems with the hardware or software. You will never be able to produce good-quality stereoscopic images under such conditions.

Accommodation-Convergence Relationship

Recall from Chapter 1 that in viewing a stereoscopic image, the eyes' axes converge as if portions of the image are at different distances, but they remain focused (accommodated) on the plane of the screen. This is the most significant respect in which an electrostereoscopic display differs from the way we see objects in the visual world. When your eyes converge on your finger, they also focus on the finger. When the eyes converge on the background, they also focus on the background. This accommodation-convergence correspondence is a habitual response learned by everyone. However, looking at an electrostereoscopic image results in a disruption of this relationship.

Because the action of the muscles controlling convergence and the muscles controlling focusing is a departure from their habitual relationship, some people experience an unpleasant sensation when looking at stereoscopic images, especially images with large values of parallax. Experience teaches that it's best to use the lowest values of parallax compatible with a good depth effect in order to minimize the breakdown of the accommodation-convergence relationship. Low parallax values help to reduce viewer discomfort.

To clarify this concept, note that when image points have zero parallax, the eyes are both converged and accommodated on the plane of the display screen. This is the only case in which the relationship between accommodation and convergence is preserved in viewing a projected planostereoscopic display. (An electronic display screen may be considered to produce a rear-projected image, in contrast to a stereoscope display, which places the two

1/2 "

L [ ] R

18 inches

→| |←1.5°

views in separate optical systems.) Accordingly, low values of parallax reduce the breakdown of accommodation-convergence, and large values of parallax exacerbate it. This subject will be discussed in greater detail in the next section.

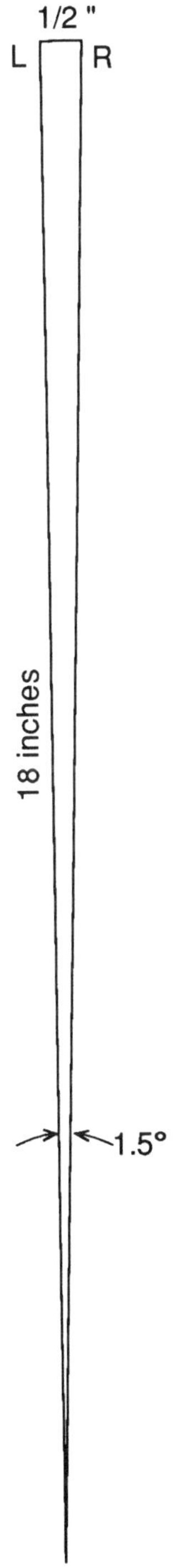

Control of Parallax

The goal in creating stereoscopic images is to provide the maximum depth effect with the lowest values of parallax. This is accomplished in part by reducing the interaxial separation. If the composition allows, it is best to place the principal object(s) at or near the plane of the screen, or to place the middle of an object at zero parallax so that approximately half of the parallax values are positive and half negative.

As a rule, parallax values should not exceed 1.5° [VALY62]. That's half an inch, or 12 millimeters, of positive or negative parallax for images to be viewed from the usual workstation distance of a foot and a half (Figure 2.5). However, this rule was made to be broken, so let your eyes be your guide. Some images require less parallax for comfortable viewing, and you may find that it's possible to exceed this recommendation for viewer-space effects.

Expressing parallax in terms of angular measure directly relates it to retinal disparity. For example, a third of an inch of parallax produces the same value of retinal disparity at three feet as two-thirds of an inch of parallax at six feet. This concept is quantified in Chapter 5. Keep in mind the distance of viewers from the screen when composing a stereoscopic image.

Cross Talk

As discussed in Chapters 4 and 6, there are two opportunities for cross talk (ghosting, or *leakage*) in an electronic stereoscopic display: departures from the ideal shutter in the eyewear, and CRT phosphor afterglow [LIPT87].

The perception of ghosting varies with the brightness of the image, color, and—most importantly—parallax and image contrast. Images with large values of parallax will have more ghosting than images with low parallax. High-contrast images, such as black lines on a white background, will show the most ghosting. Given the present state of the art of monitors and their display tubes, the green phosphor has the longest afterglow and produces the most ghosting. If an image is producing ghosting problems, try reducing the amount of green in the image.

Figure 2.5
Maximum parallax. Left and right image points should be separated by half an inch when viewed from a distance of 18 inches, shown here at almost half scale. If you're farther from the screen, more distance between the image points may be acceptable, but remember the 1.5° rule.

Curing the Problems

The breakdown of the accommodation-convergence relationship and ghost images produced by cross talk can detract from the enjoyment of a stereoscopic image. Fortunately, the cure for both problems is the same: use the lowest values of parallax needed to produce a good stereoscopic effect. As corresponding points are moved closer together, accommodation and convergence approach their usual relationship, and the ghost image also diminishes.

2.4 Composition

Although stereopsis adds another depth sense to a display, it is arguable whether this addition makes the display more realistic. Stereopsis adds information in a form that is both aesthetically pleasing and useful. However, a stereoscopic image that is isomorphic or visually equivalent with the real world may be uncomfortable to view and maybe of questionable value to the scientist, engineer, technician, and artist.

A stereoscopic image may depart from isomorphism with the visual world in several ways. Three of these ways, which involve psychophysics, the psychology of depth perception, and geometrical or optical considerations, respectively, are breakdown of accommodation and convergence, screen-surround conflicting cues, and orthostereoscopic conditions.

Accommodation-Convergence Breakdown

Many individuals, especially children, don't have any problem with the inconsistency of accommodation and convergence. With practice (i.e., with no conscious effort), many people become used to viewing stereoscopic images.

The breakdown of the psychophysical relationship between accommodation and convergence is a condition that users, software designers, and stereovideographers must keep in mind because a planostereoscopic display can never overcome this artifact. The problem is most pronounced with small screens (only a foot or two across) viewed at close distances (a foot and a half to five feet), which are characteristic of electrostereoscopic workstation displays. Large-screen displays at greater distances may be viewed with less effort, and evidence indicates that the breakdown of accommodation and convergence is less severe in this case [INOU90]. When the eyes are accommodated for distances of many yards, the departure from customary convergence may not be troublesome. Viewers of stereoscopic motion pictures employing gigantic positive parallax, typically found at theme parks, may have a surprisingly agreeable response.

Accommodation and convergence don't provide the mind with depth information [KAUF74], a statement at odds with the belief of some stereoscopists. That convergence provides any depth information has not been clearly demonstrated. As for accommodation, it may be the perception of blur (rather than muscles changing the shape of the eyes' lenses) that provides depth information. Therefore, the breakdown between accommodation and convergence probably doesn't result in a conflict of depth cues. Although it may seem like a fine distinction, the conflict is actually based on a departure from the habitual relationship of two sets of neural pathways and the muscles they control.

The accommodation-convergence relationship is learned; it is not a physiological given. People who are experienced at looking at stereoscopic images may no longer have the usual accommodation-convergence response. This can be a disadvantage, because stereoscopic images composed by such people may exceed the limits of what is comfortable for untrained people. Bear this in mind, and avoid composing images that tax other people's visual systems. If you're creating images for presentations, it's a good idea to try them out on people who are not trained in fusing stereo pairs.

Screen Surround

The vertical and horizontal edges of the display screen are called the surround. In the composition of a stereoscopic image there is a basic decision to be made: where to position the scene or the object with respect to the screen surface, or stereo window.

It's best if the surround doesn't cut off a portion of an object with negative parallax (i.e., in viewer space). If the surround cuts off or touches an object in viewer space, there will be a conflict of depth cues that many people find objectionable (Figure 2.6). Because this anomaly never occurs in the visual world, we lack the vocabulary to describe it; so people typically call the effect "blurry" or "out of focus." For many people this conflict of cues results in the image being pulled back into CRT space, despite the fact that it has negative parallax [LIPT82].

One cue, interposition, tells you that the object must be behind the window since it's cut off by the surround. The other cue, the stereoscopic cue provided by parallax, tells you that the object is in front of the window. Experience tells us that when we are looking through a window, objects cannot be between the window and us. The stereo illusion is convincing only if the object is not occluded by a portion of the window (Colorplate 2.1).

Objects touching the horizontal portions of the surround, the top and the bottom, are less of a problem than objects touching the vertical portions of the surround. Objects can be hung from or cut off by the horizontal portions of the surround and still look acceptable. The vertical portion of the surround leads to problems because of the paradoxical stereoscopic window effect. When looking outdoors through the left edge of a real window, the right eye sees more of the outdoor image at the edge than the left eye sees. The same holds true for the left eye at the right edge of the window. The difficulty in stereoscopic imaging arises from the fact

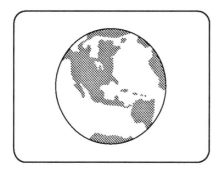

Figure 2.6

Interposition and parallax cues. There is no conflict of cues in the top drawing. The globe, in viewer space with negative parallax, is centered within the screen surround. The lower drawing shows the globe cut off by the surround. In this case, the interposition cue tells the eye/brain that the globe is behind the surround, but the stereopsis cue tells the eye/brain that the globe is in front of the surround.

that the right eye, when looking at the left edge of the screen for objects with negative parallax, will see less image, and the left eye will see more. This departure from our usual experience may be the cause of the extra difficulty associated with the perception of images at the vertical portions of the surround (Colorplate 2.2).

Viewing stereoscopic images on a large screen reduces or eliminates the conflict of cues associated with the surround. The edges of the screen may be relatively difficult to see because they're on the periphery of the visual field. Thus, the conflict of cues occurring at the surround won't be noticed. The benefits of a large screen are apparent for stereo movies projected in 70 mm using the Disney, United Artists, or IMAX system. In this regard, people at computer workstations have a more difficult task than the cinematographers working for projection on enormous screens.

Orthostereoscopy

Orthostereoscopy can affect the way in which we construct images and optics in stereo, depending on the application. Three conditions must be fulfilled for a stereoscopic image to be orthostereoscopic [SPOT53], that is, isomorphic, in terms of the stereoscopic depth cue and perspective consid-erations, with the visual world. First, images of very distant objects must cause parallel alignment of the lens axes of the viewer's eyes. Second, the distance between the perspective view points used to compute or capture the two views must be equal to the viewer's interocular distance. And third, the perspective constraint must be fulfilled. That is, the image must be viewed from a distance (V) that is equal to the product of the image magnification (M) of the object and the focal length (f) of the lens (Figure 2.7):

$$V = Mf$$

The last constraint is well known to lens designers and photographers [LIPT84a] and is true for all images photographically created or computer-generated using a camera model, as described in the next two sections.

To fulfill the first condition, the parallax values of distant image points would have to be equal to the interpupillary distance, and that's much too large for comfortable viewing. Viewing distant objects in the real world will produce parallel lens axes, but for a workstation screen the resulting breakdown of accommodation and convergence would be uncomfortable.

If we were to try to fulfill the second condition by specifying the distance between perspective views to be equal to the distance between the eyes, for many applications (such as molecular modeling or aerial mapping) the result would be useless. Molecules are very small, requiring a perspective separation computed on the order of the size of atoms, and aerial mapping often requires a stereo base of hundreds or thousands of feet. Moreover, the distance between human eyes, or the interocular distance, varies between 50 and 75 mm. A view computed for one set of eyes wouldn't produce an orthostereoscopic result for most other eyes.

If we were to take the third condition, the perspective constraint, seriously, viewers would have to change their distance from the screen for

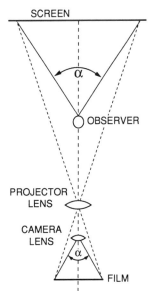

Figure 2.7

$V = Mf$. When this condition is fulfilled, the angle α subtended by the projected image of the object at the observer's eye equals that of the object subtended by the camera's lens on the film.

every new object that appeared. Nobody is going to do this. People have learned to make a mental adjustment for images produced with wide-angle or telephoto lenses.

In general, the orthostereoscopic condition is needed only for specialized photogrammetric and virtual-reality applications, where it is important to have an isometric image of the world.

2.5 More Notes on Composition

Convergence

Historically, the term *convergence* has been applied to the process used to achieve the zero parallax condition for stereo imaging. The use of the term in this context creates confusion with regard to the physiological process of convergence, requiring rotation of the eyes. Thus, we need to distinguish physiological and stereoscopic convergence.

A further complication arises from this usage. A portion of the image is said to be "converged" when homologous points coincide; however, the term *plane of convergence* has been used for the zero parallax condition, which supposedly exists for all image points located on a plane at a given distance (the distance on which the camera is converged). Unfortunately, the locus of such points will not be a plane if rotation is used for convergence [SPOT53]. Rather, it will be a complex but continuous surface.

Object Placement

For compositions resembling scenes in the real world, most of the stereoscopic image should lie within CRT space. Imagine the following southwestern scene: a cactus in the foreground, a horse in the middle distance, and mesas at a great distance (Figure 2.8), with the cactus touching the bottom edge of the screen surround. The cactus should be given zero parallax where it touches. To give it negative parallax would lead to a conflict of cues, because the stereopsis cue would indicate that the image is in front of the screen while the interposition cue would indicate it is behind. (In this particular example, giving the cactus a little negative parallax, placing it in

Figure 2.8
Parallax values. The cactus is at the plane of the screen; the horse and the mesas have positive parallax.

viewer space, wouldn't hurt because the conflict of cues for the horizontal edge of the surround is relatively benign.) The most distant objects in the scene, the mesas, will have the maximum positive parallax. Everything else in the scene has between zero parallax and the maximum value of positive parallax.

On the other hand, many computer-generated images involve a single object, for example, a molecule or a machine part. For this class of compositions, try placing the center of the object in the plane of the screen, with some of the object in viewer space and some of it in CRT space. This tends to reduce the breakdown of accommodation and convergence. Where you place the object is also an aesthetic decision.

The most important variable to manipulate for strong off-screen effects is not parallax but rather the perspective cue. It is the signature of the amateur to use huge parallax values to obtain strong viewer-space effects. Stressing perspective is a better approach, and it's accomplished by using a wide-angle view, at a bare minimum of 40° horizontal, and by scaling the object to fill the screen. This will be discussed in the following section.

Viewer Distance

The farther the observer is from the screen, the greater is the allowable parallax, since the important entity is not the distance between homologous points but the convergence angle of the eyes required to fuse the points. A given value of screen parallax over a greater distance produces a lower value of retinal disparity.

As indicated in Chapter 5, a stereoscopic image looks more three-dimensional or deeper the farther the observer is from the display screen. Distant objects look farther away, and objects in viewer space look closer. The subject is treated in more detail in [LIPT82] and [SPOT53], but, simply put, the proportional distances of parts of an image remain constant for any viewer distance. An object that looks like it's one foot from you if you're two feet from the screen will look like it's two feet from you when you move to a position four feet from the screen. Therefore, off-screen effects will be exaggerated for more distant viewers.

Images of objects may, for some compositions, have a maximum value of negative parallax larger than that acceptable for positive parallax. Objects can have large values of negative parallax and be cut off by the vertical surround if they're held on the screen for a short period of time. Objects racing swiftly off or onto the screen can pass through the vertical surround without being difficult to view. Only experiments will tell you what works best.

2.6 Software Tips

The stereoscopic effect should be under the control of the software designer or the user. There are two major factors controlling the strength of the effect: the stereoscopic component, or parallax, and the extrastereoscopic component, that is, all the other depth cues (linear perspective, interposition, etc.). Of the latter, the most important cue for stereoscopic purposes is perspective [MACA54], which is determined by the distance of the object from the camera (for computer graphics we'll assume a camera model).

The perspective cue weights or scales the stereoscopic cue. If there is a strong perspective cue produced by a wide-angle perspective viewpoint, there will be a strong stereoscopic effect. For the image size to be maintained in a wide-angle view, the object must be brought closer to the camera. By this means perspective is stressed.

Given a strong stereoscopic effect, the values of parallax may be lowered by reducing the interaxial distance. The stereopsis depth cue may be reduced in this case, because the extrastereoscopic depth cue of perspective has been strengthened. The reduction in parallax will diminish both the breakdown of convergence and accommodation and ghosting, producing an image more pleasing to view.

Here's more advice for software developers dealing with images of single objects centered in the screen: when the image initially appears on the screen, the image points at the center of the object should default to zero parallax; that is, the stereo window should slice the object approximately in half. Moreover, in this way zero parallax is maintained within the object even during zooming or changing object size. Some people don't like this notion. They think that the placement of the stereo window should be in front of the object when it is far away (small) and behind the object when it is close (big), to simulate what they suppose happens in the visual world. But in the visual world the eyes remain converged on the object as it changes distance; hence, retinal disparity, the counterpart of screen parallax, remains constant.

The depth cue of relative size is sufficient to make the object appear to be near or far, or changing in distance, and often the objective is only to scale the image to make it a convenient size. Its center should remain at the plane of the screen for easiest viewing.

Proposed Standard

Here is a proposed standard for stereoscopic software to help users control the two basic variables of stereoscopic composition: interaxial distance (t_c) and parallax. The arrow keys **N, S, E,** and **W** should be used to control these functions (Figure 2.9). **N** and **S** are devoted to the interaxial setting — **N** for increasing the distance between camera axes and **S** for decreasing the distance.

For controlling parallax there are two possible ways to change the relative direction of the image fields, and it doesn't matter which is assigned to the **E** or **W** key. That's because the image fields approach each other until zero parallax is reached, and they pull away from each other after passing through zero parallax.

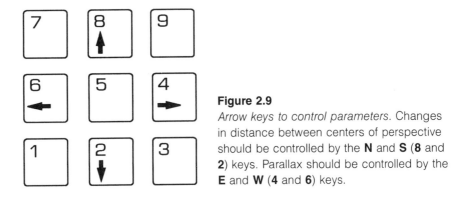

Figure 2.9
Arrow keys to control parameters. Changes in distance between centers of perspective should be controlled by the **N** and **S** (**8** and **2**) keys. Parallax should be controlled by the **E** and **W** (**4** and **6**) keys.

A helpful touch is to maintain zero parallax when the interaxial distance is varied. In this way the user can concentrate on the strength of the stereoscopic effect without having to be involved in monitoring the position of the stereo window and working the parallax keys. This standard can be menu-driven by the input device being employed to manipulate stereo images.

The Camera Model

The geometry of the basic algorithm for producing computer-generated electrostereoscopic images is illustrated in Figure 2.10, showing left and right cameras located in data space. Their interaxial separation is given by t_c. The camera lens axes are parallel in the z direction. The distance from the cameras to the object of interest is d_o.

The two cameras use lenses whose focal lengths are identical. Short-focal-length lenses produce a wide-angle view, and long-focal-length lenses

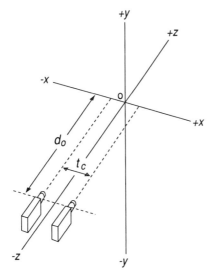

Figure 2.10
Distortion-free images. The camera axes (or the axes of the centers of perspective) must be parallel.

produce a narrow-angle view. For example, wide-angle lenses for 35 mm photography generally include anything under 50 mm, and long-focal-length or telephoto lenses are generally anything above 50 mm.

This parallel-lens-axes algorithm for stereoscopic computer-generated images uses two camera viewpoints, with parallel lens axes, set some distance t_c apart. Both viewpoints, or camera lenses, have the same angle of view. The degree to which the images are horizontally shifted or translated is more than a matter of taste. Earlier the advantages of creating images with low parallax were explained. We must think of using the stereo window to produce the best compromise of parallax values for the entire image.

The Parallax Factor

The use of small t_c values and a wide angle of view, at least 40° horizontal, is suggested. To understand why, consider the relationship

$$P_m = \frac{M f_c t_c}{d_o} - \frac{M f_c t_c}{d_m}$$

This is the depth range equation used by stereographers, describing how the maximum value of parallax (P_m) changes with changes in the camera setup. Imagine that with horizontal image translation, the parallax value of an object at distance d_o becomes zero. We then say we have achieved zero parallax at the distance d_o. An object at some maximum distance d_m will now have a parallax value of P_m.

The aim is to produce the strongest stereoscopic effect without exceeding the maximum of 1.5° of parallax. The form of this equation is helpful in clarifying this. The value of magnification (M) will change the value of P_m. For example, the image seen on a big screen will have greater parallax than the image seen on a small screen. A 24-inch-wide monitor will have twice the parallax of a 12-inch monitor, all other factors being equal. Reducing t_c will reduce the value of screen parallax. Also note that reducing f_c (the lens focal length) or using a wide-angle lens also reduces P_m.

The most important factor controlling the stereoscopic effect is the distance t_c between the two camera lenses. The greater t_c, the greater the parallax values and the greater the stereoscopic depth cue. The converse is also true. If we are looking at objects that are very close to the camera— say, coins, insects, or small objects—t_c may be low and still produce a strong stereoscopic effect. On the other hand, if we're looking at distant hills, t_c may have to be hundreds of meters in order to produce any sort of stereoscopic effect. So changing t_c is a way to control the depth of a stereoscopic image.

Small values of t_c can be employed in conjunction with a wide angle of view (low f_c). In terms of perspective, the relative juxtaposition of the close portion of the object and the distant portion of the object, for the wide-angle point of view, will be exaggerated. It is this perspective exaggeration that scales the stereoscopic depth effect. With the use of wide-angle views, t_c can be reduced [LIPT82].

This parallel-lens-axes approach is particularly important when wide-angle views are used for objects that are close. In this case, if we used rotation, the geometric distortion would exacerbate the generation of spurious vertical parallax.

Remember, the strength of the stereoscopic effect is controlled by the interaxial distance t_c, which controls the parallax values of the object or scene. Once the interaxial setting is established, horizontal image translation is used to control the zero parallax position of the images; this is often achieved in software by moving two portions of the buffer relative to each other.

In developing user-friendly software, consideration should be given to the fact that when an image is rendered, its location relative to the stereo window should be based on ease in viewing. For a single-object image, the stereo window should slice the middle of the object. If this suggestion is not followed, the user will have a difficult time because of the breakdown of accommodation and convergence, and because of the concomitant cross talk.

3

3D Hardcopy

Shaun Love

3.1 Introduction

This chapter treats the problems of producing hardcopy representations of computer-generated three-dimensional images. The possible approaches to producing three-dimensional hardcopy may be divided into the general categories shown in Figure 3.1. In order to avoid omissions, a rather broad view is taken of what might be considered hardcopy.

All of the currently viable methods for generating digital images, including holography, make use of stereo pairs or slices. At the highest level, these techniques may be divided between "unaided viewing," or autostereoscopic techniques, and "aided viewing," in which some type of viewing apparatus is needed to deliver the correct image to an observer. Autostereoscopic techniques may be further divided into holographic and nonholographic categories. Techniques requiring viewing apparatus are divided into static and dynamic images. Dynamic images allow the introduction of movement into the 3D display.

3.2 Printed Stereo Pairs

The most commonly used method for providing a viewer with an image containing depth is the stereo pair. An ordinary photograph shows how a scene appears from a particular vantage point. Presenting the viewer with twin images, recorded from slightly different positions, provides most of the binocular depth cues available when the scene is viewed directly. For these cues to be effective, however, it is necessary to restrict each image to the proper eye.

Free Viewing

An extremely simple method for providing a stereo image is the technique of *free viewing* [FERW82], described in Chapter 1. Free viewing, which may be subdivided into the techniques of parallel and cross viewing, requires no optical aids.

Stereoscopes

The Wheatstone and Brewster stereoscopes use mirrors and prisms, respectively, to present each eye with one of the stereo images ([OKOS76],

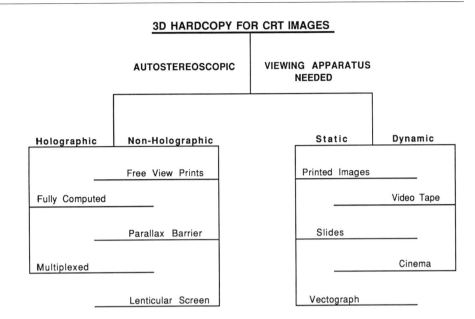

Figure 3.1
3D hardcopy
categories.

[LANE82], [BREW56]). The addition of these optical elements changes the required focal length of the lens in the viewer's eye to bring it more in line with the point of convergence. An additional benefit of the stereoscope is the elimination of peripheral distractions. Developed in the first half of the 1800s, stereoscopes were a popular means of displaying stereo photographs.

Anaglyphs

Anaglyphs superimpose the left- and right-eye views and use complementary colors to encode them. The viewer wears glasses with differently colored filters to separate the images. Since the images are superimposed, there is no restriction on parallax. The subject can be centered at the image plane and extend both away from the viewer as well as forward into viewer space. However, since viewing is through colored filters, a true color image is not possible. For computer-generated images, filters can be chosen to match the ink of a printer or plotter. In addition to generating printed images, this technique was used for some of the 3D movies that were popular in the 1950s [LIPT82]. The major drawbacks to this method are the false coloration that it produces and the severe discomfort experienced by many viewers.

3.3 Transparencies

A convenient method for displaying stereo images to a large group is the projection of transparencies [FERW82]. Separation of superimposed images is done with color filters, as in the case of anaglyphs, but the use of polarized light makes full-color images possible. Placing filters in front of two projector lenses creates a pair of stereo images that can be polarized in orthogonal directions. A third filter in front of the viewer's eye, when held in one orientation, allows one of the images to be seen and blocks the other. Rotating the filter 90° allows the second image to be seen while blocking the first. For stereo viewing, inexpensive glasses made from polarized material allow each eye to see only the appropriate image. Without the glasses, both images are always visible, and even with the glasses in place, the opposite-eye image can often be detected. This ghosting is due to inefficiency of the polarizers, imperfect filter orientation, or interference with the polarization by the projection screen. An expensive metallic screen is normally used to preserve polarization, though a much cheaper version can be made using aluminum spray paint [WIXS89]. This method can be used to present stereo movies as well as still slides simply by changing the type of projectors being used.

There are problems with the projection of transparencies that can cause distortion to the 3D image. The amount of depth perceived is a function of the viewer's distance from the display screen. When there are several viewers looking at the same display, this distance will vary, and so will each person's perception of depth. Another problem is that the projected images must be properly registered by the projectionist. If two projectors are used, there may be vertical misalignment or keystoning of the stereo images, which will make them nearly impossible to fuse [HODG88c]. In addition, panning, which is the horizontal movement of one image relative to the other, will affect perceived depth.

3.4 Vectograph

Another method that uses polarization is the Polaroid vectograph, a clear plastic laminate of two sheets of polyvinyl alcohol that can be used to record a stereo pair on a single sheet [LAND40]. The polyvinyl alcohol sheets are the transparent base material and, when dyed with iodine, become polarizing filters. In a vectograph, two sheets of the stretched, unstained material are laminated together with their transmission axes perpendicular. If iodine-based ink is used to draw an image on one side of this composite sheet, the area covered by ink will polarize light passing through it and all the unpainted areas will leave the light unpolarized. Drawing on the opposite side will also produce an image in polarized light but with the polarization oriented perpendicular relative to the first image.

For viewing stereo images, the left-eye perspective is plotted on the front side of the material, as shown in Figure 3.2, and the right-eye perspective is plotted on the back side. Images can be drawn using a flatbed plotter, or, for higher resolution and continuous tones, a dye transfer process employing matrix film can be used. However they are produced, care must be taken to ensure proper registration of the stereo images. It is also possible to make

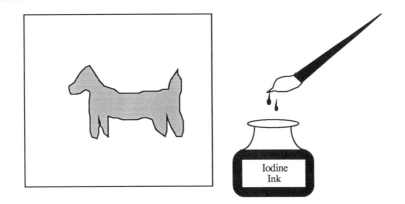

Figure 3.2
Vectograph
image
generation.

full-color vectographs. Color separation methods are applied to the two full-color images in order to separate eight CMYK primary-color images, which are then drawn in tinted ink.

The 3D image is viewed through polarized glasses. The way in which the left-eye image would appear to each eye is shown in Figure 3.3. It is also possible to display vectographs to large groups by projecting the polarized

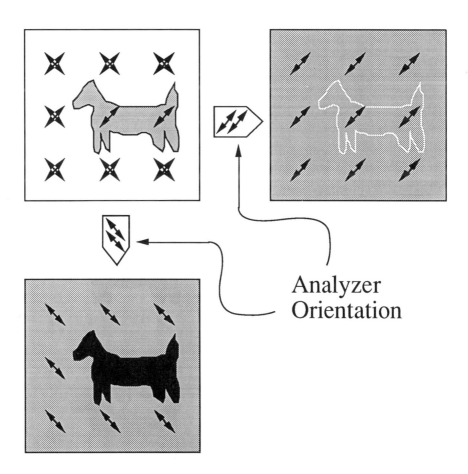

Figure 3.3
Viewing a
vectograph.

images onto a metallic screen. This is done using a single slide projector, or even an ordinary overhead projector will do. Using only one projector eliminates problems of stereo image alignment at presentation time.

3.5 Video

An extremely simple method for producing a stereo pair from a moving image utilizes the Pulfrich effect ([PULF22], [SHIM88]). A neutral-density filter placed over one eye causes a delay in the time the image is processed by the brain. With only one eye covered this way, the images processed from each eye are from different points in time. If objects in the scene are in motion, they will appear to be in a different place to each eye, and depth is perceived. Note that the perception of depth is dependent upon motion. If an object is stationary, it will appear the same to both eyes, and no depth perception results. Objects in motion will appear to be either in front of or behind stationary objects, depending on the direction of their travel and on which of the viewer's eyes is covered. This can easily be incorporated in computer-generated animations and stored on videotape. A 3D broadcast using this technique was aired as the halftime show of the 1989 Super Bowl and more recently as part of a Rolling Stones concert.

3.6 Selector Screen Methods

With the exception of free viewing, the preceding technologies all require the viewer to use some special device to separate the stereo images. Autostereoscopic displays perform the image separation for the viewer and are clearly more convenient. There are three common methods of autostereoscopic display, which differ in the way the stereo images are separated for presentation to each eye. Using diffraction, refraction, and occlusion, respectively, they are holographic, lenticular sheet, and barrier strip displays. Lenticular sheet and barrier displays share many characteristics and are known as selector screen display methods [OKOS76]. This name is indicative of the fact that a screen placed between the viewer and the printed image selects what is displayed to each eye.

Parallax Barrier

Dating back to the early 1900s [IVES03], a parallax barrier consists of a series of fine vertical slits in an otherwise opaque barrier. Placing such a barrier a slight distance in front of a sheet of photographic film causes any image projected through the barrier to be recorded on the film as a series of fine strips. If the projector were removed and the viewer's eye located in that position, the strips formed by the projected image would be visible. Provided that the slits occur at an adequate sampling frequency, image quality is maintained although brightness is significantly reduced. Regions of the film plane that were shaded from the first projected image can be used to record additional images. By moving the projector slightly to the side, the second stereo image is recorded where it can be seen by the viewer's other eye. The barrier ensures that the strips from each image will be visible only from the proper viewing direction so that each of the viewer's eyes is provided with a different view.

The viewer depicted in Figure 3.4 is correctly positioned to see an *orthoscopic* or normal image. If this viewer were to move slightly to the side, the image seen by the left eye would be the one intended for the right eye, and vice versa. With each of the stereo views presented to the wrong eye, the viewer sees a *pseudoscopic* image, in which depth relationships are reversed. If the viewer continues to move in the same direction, the views will alternate between an orthoscopic image and a pseudoscopic image.

If a display has only two images behind each slit, the number of pseudoscopic and orthoscopic views will be equal. Reducing the size of the slits or increasing the distance between them allows a greater number of images to be recorded behind them, providing several advantages. The most obvious effect is that the image is now panoramic. If the viewing position is changed, a different orthoscopic stereo image will be visible that is correct for the new viewing position. Another advantage is that the ratio of orthoscopic to pseudoscopic views is increased, meaning that a viewer positioned at random has a lower risk of seeing a pseudoscopic image.

Parallax barrier displays can present very high-quality images. In addition, because they are autostereoscopic, they are very convenient for the viewer. The main drawback to the parallax barrier technique is that since so much light is blocked by the barrier, the image is often dim, and it is

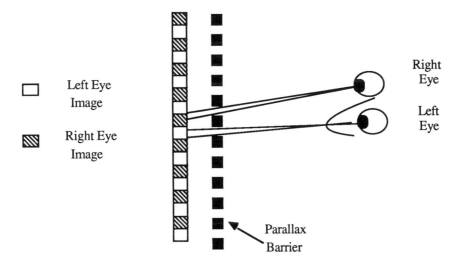

Figure 3.4
Viewing a parallax barrier display.

only practical to display backlit transparencies such as Cibachrome display transparency film.

Lenticular Sheets

A lenticular sheet consists of a series of cylindrical lenses, typically rolled out of plastic. The sheet is designed so that parallel light entering the front of the sheet will be focused onto strips on the flat rear surface. A three-dimensional display is produced by placing image strips on the rear surface behind each of these lenses as shown in Figure 3.5. Each eye's line of sight is focused by the lenses onto a different part of these image strips, giving a stereo pair.

Barrier displays are often considered superior to lenticular ones, but the quality of a lenticular display depends on the uniformity and quality of the lenses. Large, high-quality lens sheets can be produced, though this is a difficult task, especially in quantity, and the sheets are subject to shrinkage during fabrication. They can also be improved through the use of elliptical rather than cylindrical lenses [BÖRN87]. Lenticular displays do offer at least one significant advantage over barrier methods. Since they use refraction rather than occlusion, image brightness is superior. Lenticular displays have no special lighting requirements, making them more portable and useful with images recorded on any medium, not just transparencies. This fact, combined with the ability to fabricate lens sheets inexpensively from plastic, accounts for the popularity of lenticular over barrier strip displays. Three-dimensional lenticular sheet pictures have been published in magazines (e.g., the April 7, 1964, issue of *Look*) and are often used for 3D postcards, greeting cards, and novelty items. This is also the technique used to create 3D photographs with the Nimslo and Nishika cameras.

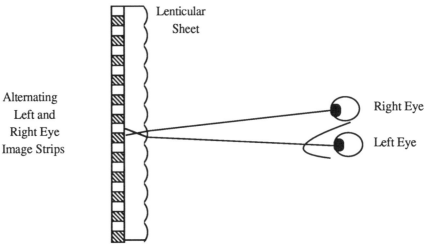

Figure 3.5
Viewing a lenticular stereogram.

With selector screen displays, since the image is physically distinct from the selector, it is not necessary that the images be static. This suggests several possible applications, such as lenticular CRTs or even a lenticular projection screen for an autostereoscopic theater. Such a theater was demonstrated at the Soviet pavilion at Expo '70 in Osaka ([OKOS76], [COND84]).

3.7 Holography

Holography is a form of lensless photography that was introduced by Gabor in 1948 [GABO48]. It is based on film recording, but it does not record the image in the same way as traditional photography, and the negative it produces will appear blurred or smudged. When properly illuminated, holograms can produce 3D images that are virtually indistinguishable from real objects. By shifting position, the viewer can look around or over objects in the foreground to see what is behind them. Even though the hologram is recorded on black-and-white film, the reconstructed image is in color—usually monochrome, although full color is possible.

Many different types of holograms can be made, incorporating variations in both viewing conditions and fabrication techniques. Perhaps the best known is the white light reflection hologram. These holograms are found on credit cards, magazine covers, and record packaging, and they are even given away as premiums in cereal boxes. Because of its unique capabilities, efforts to use holography to display computer-generated images have been pursued since the 1960s. These can be placed into two broad categories: fully computed and multiplexed.

Fully Computed Holograms

To create a fully computed hologram, one calculates and draws the fringe pattern required to reconstruct the desired image. Though these holograms are conceptually simple and potentially quite effective, their success has been limited due to the capabilities of current writing technology. Modern holographic films such as Agfa-Gevaert 8E75HD claim a resolution of greater than 2,000 lines/millimeter with a grain size of 30 nm (compared with 185 lines/millimeter for Agfapan-25 professional, a film used in common photography) [YZUE83]. It is beyond the capacity of any current output device to exploit this bandwidth or storage density. Efforts to approach this resolution began with the use of plotters to make large-scale drawings, which could then be photoreduced. Surprisingly, for very simple objects such as stick figures, extreme demagnification is not needed to obtain a discernible, although noisy, image [MARS84]. The use of photoreduction has largely been replaced by methods of writing the pattern directly onto a substrate. Electron-beam writers can obtain submicron-level resolution but record a binary pattern ([ARNO85], [FREY83]). The substrate used is chrome on glass, and any point is either transparent or opaque. A gray scale can be introduced by halftoning, but only with a loss of resolution.

The low resolution of current plotting devices leads to several problems, of which image degradation is only one. There are also problems in trying to view a reconstructed image. Widely spaced fringes produce only a small separation of diffraction orders, so the reconstructed image (first diffraction order) will be only slightly removed from the bright light that passes directly through the hologram (zeroth order). This can make viewing difficult, and large images from adjacent orders may even overlap. In addition, there is usually low diffraction efficiency, resulting in a dim image, though brightness can be improved by making a phase hologram. At present, fringe writing capabilities are still very poor.

Holographic Stereograms

The primary difficulty with fringe writing is the inability of output devices to draw the pattern at the requisite resolution. To overcome this limitation, multiplex holograms revert to the proven technique of recording the hologram by laser.

If a hologram is covered by a sheet of cardboard with a peephole cut in it, the entire object may still be seen through the hole. If the peephole is small enough, only one perspective view is available. To produce a holographic stereogram, the hologram is subdivided into very many such peepholes. For each peephole, the perspective view it would afford is calculated, rendered on a CRT, and photographed. Finally, a laser is used to record a holographic image of the photo into the peephole or subhologram (see Figure 3.6). When the finished hologram is illuminated, each eye looks through a different peephole and the viewer thus is presented with a stereo pair. Breakdown of the accommodation-convergence relationship results, as with any stereo pair technology [BENT82].

To limit image flipping, the extent of each subhologram should be no greater than the diameter of the pupil of the eye. For most people 3 mm is adequate. It is useful to simplify calculations by eliminating vertical parallax, so rather than using 3 mm square peepholes, the hologram is usually divided into narrow vertical strips, with each recording the view from only one height. Such a hologram has the added advantage of being viewable in white light [BENT69].

As in standard photography, full color is achieved through the combination of three images, each in a different primary color. Rainbow holograms are a variation of monochrome; they produce differently colored versions of the same image, but there is no color mixing. They can be reconstructed in white light because of their lack of vertical parallax. To make a rainbow transmission hologram, a horizontal slit is used in the recording process. If a monochromatic light source is used to reconstruct an image, the image of the slit will be visible. If a two-color source is used, the hologram will

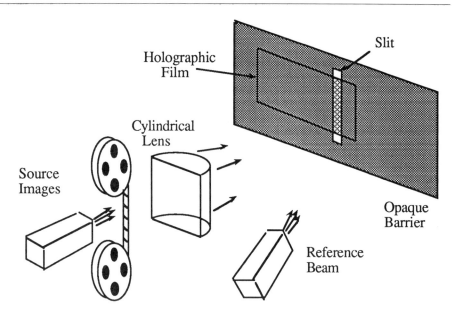

Figure 3.6
Recording a holographic stereogram.

diffract the colors at different angles, and two monochrome images of the slit can be seen. With a white light source, a continuous range of slits is smeared in the vertical direction, giving the familiar rainbow appearance.

It is possible to produce a full-color holographic stereogram by super-positioning three rainbow holograms in each vertical strip [MOLT84]. To accomplish this, a color image is separated into its RGB components. By using different angles of the reference beam, images are recorded so that the desired colors from each will be reconstructed at the same vertical position. When viewed at the proper height, the RGB images blend to give the desired coloration. Head position is restricted to this one level since incorrect coloration will be seen at any other.

Holographic Cinema

The ability to present autostereoscopic 3D movies excited the imagination long before moviegoers watched the robot R2D2 deliver his holographic message. Although this was accomplished in Hollywood with special effects, a forty-five-second holographic motion picture was successfully demonstrated in the Soviet Union [BENT80] even before the release of the motion picture *Star Wars*. In the system developed at the Russian Cine and Photo Research Institute, the screen is an elliptical mirror. The viewer's head is positioned near one of the focal points, and a series of real holographic images is projected at the other. Due to the geometry of the mirror, light from the image striking the mirror is reflected back toward the viewer. To accommodate more than one viewer, the screen is actually a holographic optical element that functions as several superimposed elliptical mirrors.

Fabrication of such a screen is difficult, and the number of viewers is limited to four.

3.8 Display Characteristics

It is apparent that there is no perfect solution for all 3D applications. Each of the hardcopy methods discussed here has its advantages and disadvantages depending on the ultimate use and requirements. Except for fringe writing and the recording of multiplanar data in a multiplexed hologram, all of the viable techniques depend on the ability to record either single or multiple stereo pairs and therefore involve the conflict of accommodation and convergence. The following list presents a few of the advantages and disadvantages of each technology.

Advantages	Disadvantages
Prints	
• Cheap and easy • Mass production	• Questionable permanence of both paper and ink
Slides	
• Cheap and easy • Readily available • Easily duplicated • Useful for single-viewer or audience projection • Very long life	• Requires silver screen • Requires two projectors with polarizing filters • Registration problems or hand-mounted slides for stereo projector
Vectograph	
• Permanent mount includes polarizer, so no further registration problems • Uses single projector • Larger format available to improve resolution	• Requires metallic screen • Requires double-sided plot or dye transfer process • Ink is very corrosive • More expensive
Videotape	
• Can be broadcast • Mass production	• Moderate/low resolution • Problems with flicker
Parallax Barrier	
• Autostereoscopic • Panoramic • Large format for billboards/movie screens	• Often dim • Limited number of views

Advantages	Disadvantages
Lenticular Sheet	
• Autostereoscopic	• Lower resolution
• Panoramic	• Lens quality is usually poor
• Brighter images	
• Large format for billboards/movie screens	
Holography	
• Very high realism	• Fully computed holograms not yet practical
• Easily transported even in very large format	• Recording is difficult and requires special equipment
• Reproducible	• Replay may require monochrome light
• Can be multiplanar	• Color is very difficult

One of the most important characteristics of a stereoscopic display is unobtrusiveness. Emphasis should be on the task to be performed, not the display or hardware required to view it. With an autostereoscopic display, there is no need for additional equipment beyond the display itself. The viewer is not required to wear a helmet or special glasses, and there are no wires or tethers limiting freedom of movement. No viewer training, preparation, or even cooperation is needed to view a stereo image, nor is the audience size limited by such factors as the number of pairs of stereo glasses available. This makes autostereoscopic displays more suitable for public veiwing as it maximizes the potential number of viewers.

Another important characteristic is the flexibility to use a technology under different circumstances. Anaglyphs, for example, can readily be used either in hardcopy or with almost any color display. Many techniques traditionally thought of as hardcopy only are now being adapted to softcopy displays. A variation of the parallax barrier method used by Dimension Technologies [EICH90] to create an autostereoscopic liquid-crystal display is described in Chapter 6. Other efforts have included putting a lenticular faceplate on the front of a CRT screen. The limited success of the lenticular CRT was primarily due to the relatively low resolution then available.

Since the intent of any display system is to enhance viewer understanding or involvement, an effective display should provide the viewer with the maximum amount of information possible. Panoramic displays not only allow viewers to look around objects in the foreground but also provide the additional depth cue of motion parallax. By shifting viewpoints, even viewers who are stereoblind are able to perceive depth.

Selector screen displays are a good choice for many applications. Being autostereoscopic, they provide maximum viewer convenience while maintaining a high informational content through panoramic, full-color images. The remainder of this chapter will detail the techniques used to create these displays and will consider the benefits and problems associated with alternative display shapes.

3.9 Selector Screen Display Generation

There are a great many similarities between parallax barrier and lenticular sheet displays. In the following discussion, techniques for creating a barrier display are described, but much of what is said applies to lenticular displays as well. The focus is on barrier displays because their image quality is generally superior and, perhaps more important for casual use, anyone can have barriers made cheaply and to their own specifications.

A parallax barrier is readily made by drawing grid lines on a sheet of orthographic film with a high-resolution typesetter. Orthographic film has very high contrast and is used in the printing industry for masks to create offset printing plates [LATH79]. In ordinary commercial use, an image to be formed on a printing plate is first created as a transparent region on orthographic film. This film is used as a mask in the exposure of the photosensitive surface of a printing plate to an arc lamp. Since all other areas of the film are highly opaque, the transparent region allows the image to be burned into the surface of the plate. The extremely high contrast between the opaque and transparent areas of the film, combined with the very high resolution of the typesetter, makes this material ideal for use as a parallax barrier.

To create a barrier, a typesetter is used just as if it were an ordinary laser printer. The main differences between the two are that the typesetter has resolution up to ten times what a laser printer can achieve, and its output can be either paper or film. Since the barrier is just a set of grid lines, a drawing program can be used to produce lines with the desired thickness and spacing. Use caution when selecting which drawing program to use because some are intended for use with a laser printer only and will not generate output with a resolution higher than 300 dots per inch (dpi). To overcome this, you can use a different drawing package, or, alternatively, a very simple program to draw the grid can be written directly in the page description language. However it is produced, a barrier with forty grid lines per inch is a good starting point, though this number can be increased or reduced if desired. The width of the slit openings is quite variable and typically ranges from one-fifth to one-fifteenth of the distance between adjacent slits.

Once a suitable barrier is obtained, the problem remains of generating the composite image to mount behind it. In this composite image, the area behind each slit must be a blend of multiple views so that different images are seen at different viewer positions. One method of generating panoramic images uses a variation on a streak camera ([OKOS76], [IVES28]). In the camera depicted in Figure 3.7, the film is located immediately behind a series of fine slits that have the same spacing frequency as the selector screen. With the shutter left open, the camera is moved horizontally as the slits are swept across the film a distance equal to the space between two adjacent slits. Through each slit is recorded a continuous image containing all views from the extreme left to the extreme right positions. An advantage of this continuous image is that it provides a very wide angle of view and no limitations are placed on viewing angle or distance.

For computer generation, it is not possible to produce a composite image that is continuous. Instead, a finite number of discrete views must be merged. Just as in the case of holography, there are two general methods available to accomplish this. The traditional method is to interleave the separate views by optical means. Unlike holography, however, because the necessary resolution is much lower, it is also possible to merge these images digitally and write the composite directly to an output device.

Optical Interleaving

In the traditional approach, the images to be combined are recorded separately on film and can be previewed with a stereo viewer. The subdivision of these images into strips that blend to form the composite is accomplished by projecting them through a sampling barrier onto a sheet of unexposed film. The projection can be done using a standard photographic enlarger, a slide projector, or any other apparatus capable of focusing an image onto an easel holding the barrier and film. Only light passing through the transparent slits can reach the film. Strips from consecutive images are interleaved through either of two slightly different approaches. In one, a separator is sandwiched between the barrier and film, just as during viewing. In the other approach, the separator is omitted, and the sampling barrier is in direct contact with the film.

In the first approach, where the separator is used for both recording and viewing, the geometry of the display is constant, making the process

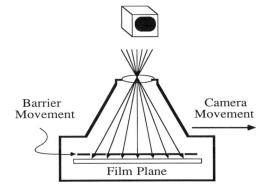

Figure 3.7
Camera for recording continuous images.

straightforward. Images are recorded with the projector located at each of the positions shown in Figure 3.8. These images will later be visible to a viewer whose eye is positioned where the projector had been. For panoramic stereo, first compute images that depict the subject from several different viewpoints. Each of these is recorded onto the composite by projecting it from its corresponding eye position, and it can subsequently be viewed from those same points. This technique is quite simple to understand, and the images to be merged are easy to compute. Simply use the location of the projector lens as the center of projection and apply a standard perspective transformation.

Despite the simplicity of the process, use of the separator during recording can cause problems. If the projector is not a point source, the image strips will suffer somewhat from spreading caused by the penumbra of the barrier shadow. Extreme reduction of the aperture of the projector lens will solve this, but at a cost of very long exposure times. In some cases, necessary exposure time has been as much as forty minutes for each view [MARS84]. To avoid both of these problems, the separator can be removed so that the barrier is in direct contact with the film. This allows only the area directly behind each slit to be exposed regardless of the lens aperture or position. This means that it is no longer necessary to move the projector to a new location for each exposure. It also means that the only way to record image strips at different positions is to physically shift the barrier relative to the film. To interleave consecutive images, the barrier must be shifted between exposures by a distance of one slit width, which may be as little as 1/600 inch. This is not trivial, but it can be done accurately by using a micrometer.

It is important to note that when the recording process is carried out without the use of a separator, the geometry of the display is altered when the separator is added for viewing. When recording is done with a separator, the image strips are spaced slightly farther apart than the slits. This is because light from the projector is diverging as it reaches the slits and continues to diverge as it passes through the separator to the film (see Figure 3.8). Without the use of a separator, the image strips are spaced exactly as far

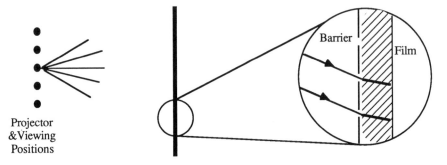

Figure 3.8
Recording and viewing with a separator.

Projector &Viewing Positions

Barrier

Film

apart as the slits. As a result, when the separator is added, light from the image strips passing through the slits will not converge but will travel in parallel planes, as shown in Figure 3.9. Each of the separate views recorded in this way should be computed as an orthographic or parallel projection. The viewer will still see an image with appropriate perspective, but this is because he or she sees small segments from many different images at once. Even though they are from different images originally, these segments blend together smoothly to give a piecewise perspective projection.

Direct Writing

For maximum speed and convenience, it is desirable to produce an image that can be displayed immediately. This makes it necessary to interlace the multiple perspective views computationally rather than optically. Calculation of an interlaced composite is actually quite simple. Merely compute views for several different viewpoints and divide them into vertical strips, which are then rearranged.

Once calculated, the interlaced image must still be printed, and a very high-resolution output device is required. A method developed jointly by the Electronic Visualization Lab, (Art)[n], and Trioptics employs a graphic arts color separation scanner to draw the interlaced color image strips [SAND89]. Several perspective views of a scene are calculated, divided into substrips, and interlaced. In the prepress industry, when a color image is prepared for printing, it is first separated into cyan, yellow, magenta, and black primaries, and an output file is generated for each. These files are taken as input by the scanner and rendered onto high-contrast, black-and- white (orthographic) film. This is the same film that is commonly used

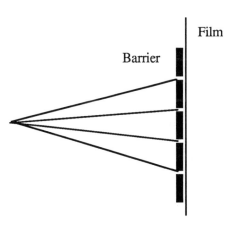
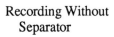
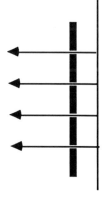

Figure 3.9
Changing display geometry by adding a separator.

Recording Without Separator   Display With Separator

to create the display barrier. The films produced by the scanner are used as masks to control the etching of printing plates, which will print these color components in four passes over the same page. Because the prepress process and printing can be costly, image proofs may be made as a means of verifying that the films are correct before the plates are etched. If a large number of identical displays is not needed, such proofs are an economical choice. Once generated, the proof is mounted on the rear surface of the separator with the line barrier on the front. Since the image strips are laid down at the same constant interval as the barrier slits, orthographic images should be calculated.

Once the quality and accuracy of the 3D image is verified, plates can be manufactured and the interlaced images mass-produced. A very important development is the ability to make copies of the display without having to mount all of them by hand. The Stealth Negative [SAND90] is created by printing the image strips on a clear, stable substrate and then turning it over and printing the line barrier in a fifth pass on the reverse side. It is of critical importance that during any printing process the four primary images be properly registered. Since registration of multiple images is a constant concern in printing, the inherent accuracy built into the printing process virtually eliminates problems in aligning the barrier with the image strips. The images produced this way have been 20 × 24 inches, though larger sizes are possible. The thickness of the substrate has varied from 0.25 inch, which accommodates thirteen to seventeen different perspective views, down to a scant 20 mils, with nine to eleven different views.

A very significant feature of these digitally interlaced displays is that the clarity and sharpness of the 3D image is enhanced substantially. This is true of both the Stealth Negative and the hand-mounted proofs. Technical applications of this work have included the AIDS virus (featured on the cover of *IEEE Computer Graphics and Applications,* November 1988), fractals, random dot stereograms with Bell Labs, and the space shuttle with NASA Ames. Artistic pieces have been shown at Feature Gallery in New York and the SIGGRAPH Art Show.

3.10 Viewing Geometry

To calculate images that will accurately depict a subject, it is necessary to determine where the sampled image strips will be visible. This requires a consideration of the display geometry. Figure 3.10 shows a barrier slit and behind it the multiple image strips visible from different viewing angles. The width of each image strip is w, which is the same as the slit width. The separator between the barrier and the image strips has thickness t and index of refraction n. The image strips are equally spaced, and a strip's location behind the slit determines the angle at which it can be viewed.

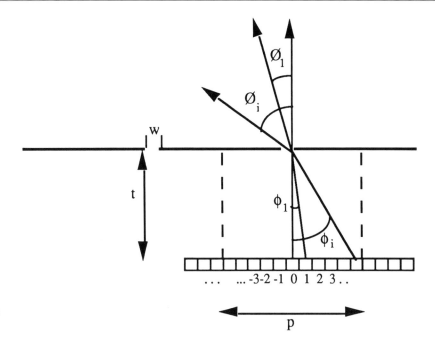

Figure 3.10
Angles at which image strips are visible.

A viewer looking through a slit from straight ahead would see strip 0. As the viewer moves to the side, the viewing angle $\varnothing$ and internal angle ϕ increase from zero, and other strips come into view. The angle from the center of the viewing slit to the center of image strip i is determined by the following equation:

$$\phi_i = \tan^{-1}\left(\frac{i\,w}{t}\right)$$

If $\varnothing_i$ is the viewing angle at which strip i is visible, then, using Snell's law,

$$\varnothing_i = \sin^{-1}(n \sin\phi_i)$$

$$= \sin^{-1}\left\{n \sin\left[\tan^{-1}\left(\frac{i\,w}{t}\right)\right]\right\}$$

It is the values of $\varnothing_i$ that are needed for the image calculations. Orthographic projection calculations require only the direction of projection, which is determined directly from $\varnothing_i$. For perspective projections, nearly any point can serve as the center of projection so long as it is in the direction indicated by $\varnothing_i$; the only other consideration is how far from the display the projector will be during the recording process.

For any display, one of the most important characteristics is the size of the primary viewing zone. Figure 3.11 shows how the image strips located

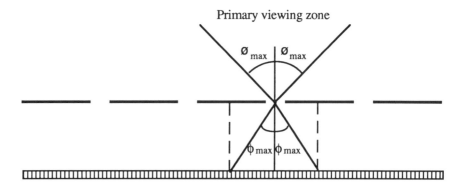

Figure 3.11
Primary viewing zone.

behind a slit will be visible at viewing angles of up to $\pm\varnothing_{max}$. To find $\varnothing_{max}$, we first find the internal angle ϕ_{max}:

$$\phi_{max} = \tan^{-1}\left(\frac{p}{2t}\right)$$

$$\varnothing_{max} = \sin^{-1}(n \sin \phi_{max})$$

$$= \sin^{-1}\left\{n \sin\left[\tan^{-1}\left(\frac{p}{2t}\right)\right]\right\}$$

At viewing angles greater than $\pm\varnothing_{max}$, the image strips that will be visible are those located behind the adjacent slits. This means that at viewing angles beyond $\pm\varnothing_{max}$, the image will begin to repeat. Displaying the same information at different locations (secondary viewing zones) increases the number of possible viewing positions.

For stereo, the number of image strips behind each slit can be as few as two, but it is generally desirable to have many more. A larger number of strips not only allows multiple stereo viewpoints but also provides the depth cue of motion parallax as the viewing position is changed. Since the images are discrete, however, a distracting feature can accompany this. As a viewer changes position so that strip i is obscured and strip $i + 1$ becomes visible, there can be a noticeable jump of the image, or "flipping," from one perspective view to the next. The magnitude of this jump is determined by the amount of parallax and can be reduced in two ways. Objects can be restricted to depths close to the display surface, where parallax is zero. The alternative is to increase the thickness of the separator so that the angular spread of the primary viewing zone is narrowed.

When considering image flipping, remember that even though the individual image strips recorded in the display are discrete, the range of viewing positions is continuous. It is certainly possible for a viewer to occupy a position where the boundary between one image strip and the next is visible. As the viewer's eye moves across this boundary between adjacent images, the first image will not be instantaneously replaced by the second but will fade out as the replacement fades in. The view provided at these transitional positions will be a weighted combination of the two consecutive images. Whether such a smooth transition is desirable or problematic depends on the parallax between corresponding points in consecutive images. If the parallax is small, a gradual fading from one image to the next is desirable, as it is nothing more than a form of antialiasing. As parallax increases, objects will begin to smear horizontally, and the image will blur as if the viewer were experiencing double vision. In the worst case, where parallax is extreme and object size is small, having images gradually come into and out of view means that the viewer will see two physically disconnected images of the object at the same time. The extent of this ghosting is dependent upon the parallax between image points in consecutive views, which is in turn dependent upon image depth.

Ghosting at image boundaries has serious implications concerning the depths of objects relative to the view screen. For dramatic effect, many viewers would like to have stereoscopic images extend in front of the screen into viewer space. Unfortunately, there is a severe price to be paid for this. The amount of parallax required to bring objects forward is substantially greater than that required to move them back. The same amount of parallax that would place a point infinitely far behind the view screen would, if reversed, bring that point only halfway between the screen and the viewer. As objects are brought out even a small distance in front of the view screen, the amount of parallax, and therefore the amount of blur or ghosting, increases substantially.

3.11 Alternative Display Shapes

Cylindrical Displays

Traditionally, displays have been almost exclusively planar. A notable exception to this is the cylindrical Cross hologram ([FUSE80], [UPAT80]). Panoramic displays provide lookaround capability, but with a planar display, the viewer's lookaround is limited since it is not possible to actually move behind the plane of the display. With a cylindrical format, the viewer is provided with maximum lookaround capability. The image can be seen from any angle, and it is possible to move completely around the display. An additional benefit is that viewers can surround the display, and so audience size may be increased. A planar display conveys information over a limited viewing range to a limited audience. A cylindrical display maximizes both.

Ordinarily, the viewer will be located at a position within the primary viewing zone of every slit of the display. If the viewer should move too close to the display, the angle between the surface normal and the viewer's line of sight to the edge of the display will exceed $\varnothing_{max}$. This means that the viewer will see the image begin to repeat itself in the secondary viewing zone. With planar displays, this can be prevented by observing a certain minimum viewing distance. For cylindrical displays, the surface normal is no longer constant, and at the extreme edges of the cylinder, the viewer's line of sight is tangent to the surface. Moving back does not solve the problem but simply changes the point of tangency. Because of this extreme viewing angle near the edges, secondary images can be avoided only if a ray coming in tangent to the barrier surface remains in the primary viewing zone. In practice this is not worthwhile, and secondary images will be visible, though steps can be taken to minimize them. Reducing the thickness of the separator widens the primary viewing zone, and the separator used for a cylinder has a thickness only about one-tenth that of separators used for flat displays. However, the problem of flipping can become quite severe because there are too few distinct images for such a wide angle of view. Increasing the distance between slits will also widen the viewing angle and will not increase flipping. Too large a separation between slits, however, will result in a display that is not only dim but has excessively coarse image resolution.

Figure 3.12 shows a cylindrical display with both primary and secondary images visible to viewers. Only the section directly in front of the viewer contains the primary image, which is necessarily smaller than what the cylinder could contain. Under any circumstances, the object size must be limited to the size of the cylinder. Although it would be possible for an object to appear to extend either in front or or behind the cylinder, it could not extend out to the side. If the cylinder were rotated, any extensions outside the cylinder would appear to be cut off.

Alcove Displays

The cylindrical format provides the maximum possible lookaround, but one of its drawbacks is a severe limitation on the size of the image it can provide. In many cases there are secondary images, and the primary one must be significantly smaller than what would be able to fit within the

Figure 3.12
Viewing a cylindrical display.

Figure 3.14
Viewing an
alcove display.

planar alcove naturally requires multiple planar segments. The number of these display panels can be as few as two and can increase without limit. Two panels would form a display similar in form to a book that is open for reading. A cylinder, of course, is the limiting case of a large number of segments. The number of panels used is a matter of preference, but the fewer the number, the less computation required.

The advantage of a two-panel display is that it requires the least amount of computation while providing the maximum portability. The two panels could be hinged, allowing the display to be closed like a book for easy transport. A disadvantage is the presence of a joint between the two panels in the center of the display. There will be a much higher likelihood of distortion at the joint than at other points of the display. Because of either faulty construction or incorrect viewing height, distortions would be apparent and centrally located. The advantage of a three-panel display is that these joints, along with their distortions, are moved somewhat off center. The number of panels and, therefore, the amount of computation required is still small, and the center panel would be identical to a planar display in both shape and orientation. In fact, the center panel could be used independently as a traditional planar display.

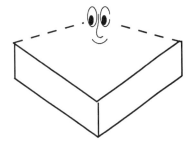

Figure 3.15
Semicircular and
piecewise planar
alcoves.

cylinder. An alternative to the cylinder is needed in order to allow incre
image size and avoid the distraction of secondary images. With a pl
display, images occupy the entire width of the display since the sur
does not fall away from the viewer. The maximum image size possible
a planar display is determined by $\varnothing_{max}$ and the viewing distance, as sh
in Figure 3.13. Making the display wider does nothing to increase this
it would only add to the secondary images. Increasing the maximum im
size and having it occupy a larger part of the viewer's visual field
be done much more effectively with an alcove format, where the disp
actually bends toward the viewer.

At MIT work has been carried out by Stephen Benton on what is ca
an alcove hologram ([BENT87, BENT88]). This hologram is illumina
from the rear and rests in a semicircular frame, as shown in Figure 3.
Like an inside-out cylindrical hologram, it produces a real image that see
to float in front of the hologram, with the alcove serving to frame it. T
hologram is composed of nearly one thousand separate perspective vie
recorded onto adjacent strips.

The alcove format increases the angle of view that can be provide
A very wide viewing angle helps to prevent peripheral distractions fro
competing for the viewer's attention. Testimony to the impact of a ve
wide angle of view comes from the decline in work on 3D displays with t
advent of Cinerama [OKOS76]. For firsthand experience of the compelli
effects of an ultrawide-angle display, attend a performance at a planetariu
or an Omnimax or Circlevision theater. Although it is possible to use a
extremely large display that is flat, to cover the largest field of visic
possible and to do so economically, an alcove format is needed.

The function of the alcove shape is to maximize image size and re
move peripheral distractions. The display need not be round to provid
this benefit [LOVE90]. It could instead consist of several flat panels joine
together in the shape of an alcove, as shown in Figure 3.15. Forming

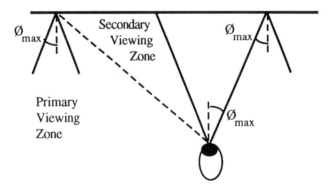

Figure 3.13
Maximum usable
image size for a
flat display.

3.12 Conclusions

As the use of stereo gains acceptance, there will be an increasing need to store 3D images in hardcopy form. A variety of technologies are available to suit different budgets and needs. Among these, slides will certainly remain a very popular medium, either individually or combined to form stereo pairs. Polarized projection of slides makes high-quality stereo images available to very large audiences. In addition to aiding the visualization of complex scientific data, selector screen displays are also being used as advertising tools in locations ranging from airport terminals to video game sales counters. Through the methods described in this chapter, parallax barrier displays can be created by nearly anyone and offer an attractive and impressive display method.

4

Visual and Perceptual Issues in Stereoscopic Color Displays

Yei-Yu Yeh

4.1 Introduction

Displays are designed to present information. In the design of any display, two primary issues are of concern: (1) image quality, for user encoding of information; and (2) display format, for efficient and accurate processing of the information. For maximization of human performance in human-machine interaction, the understanding of human capabilities and limitations is of paramount importance. At the front end, human spatial and temporal sensitivity functions dictate the display resolution for better image quality and information encoding. As the signals are encoded, human information-processing strategies have important implications for information manipulation and representation on the displays.

The purpose of this chapter is to provide a short review of visual capabilities and limitations affecting the development and application of stereoscopic visual displays. Two issues that are discussed—discordant accommodation and binocular vergence and the presence of ghost image resulting from the cross talk between the left- and right-eye views—are specific to certain implementation methods, and most other topics have direct implications for image coding on any stereoscopic color display. Additional topics are included to introduce research issues currently emerging as important in basic visual sciences. Better understanding of the fundamental aspects of stereopsis would ultimately lead to improved formats for stereoscopic color displays.

Human vision evolved to analyze a three-dimensional world, and our two eyes see the visual world from slightly different viewpoints due to their lateral separation. The difference in relative position of visual images on the retinas, lateral retinal image disparity, is an important cue in perceiving the depth of an object relative to the fixation point. When monocular depth cues are completely masked in a random-dot stereogram, lateral image disparity produces depth perception. Binocular disparity also plays an important role in shape or curvature perception. Stereopsis has been an important research field in basic visual sciences ranging from the understanding of the neurophysiological mechanisms to the investigation of interaction between the two monocular channels in visual coding.

Since Wheatstone [WHEA38] used the stereoscope to demonstrate that binocular disparity produces depth perception, many methods have been invented and implemented to simulate this natural viewing perspective. These methods include anaglyphs, with the two eye views superimposed in complementary hues and viewed through filters of complementary hues and binocular displays that use optical hardware to present the two eye views (e.g., the helmet-mounted display). Another method is to present the two eye views sequentially at a fast rate of approximately 120 Hz so that each eye receives an image at the nonflickering rate of 60 Hz, and the images from both eyes are fused. Based on the horizontal disparity in the two eye views, an image is perceived at a certain depth relative to the screen. A field-sequential stereoscopic color display typically consists of three components: a shutter device (e.g., a liquid-crystal mounted shutter composed of a π-cell and a quarter-wave retarder to provide right- or left-hand polarization), a high-resolution color monitor with a refresh rate of 120 Hz, and a pair of polarized glasses. The technical aspects of the shutters are described in Chapter 6.

There are advantages and disadvantages to the different implementation methods. Anaglyphs have the disadvantage of being incapable of presenting multicolor images. The stereoscope has the advantages of (1) presenting images at high luminance, because no polarization is required and hence transmission is not reduced, and (2) presenting stereo images with no ghosting. However, precise mechanical alignment is necessary to ensure that the background images from the two screens are aligned through reflection of the mirror for fixation. Field-sequential CRT stereoscopic displays, as described in Chapter 6, have the advantage of generating high-resolution, full-color images in real time. But a field-sequential stereoscopic color display has the disadvantage of producing perceptible ghost images when the decay rate of a phosphor (e.g., P22 green) is slow, and hence residual images remain on the screen while the sync signal is sent to switch to the other eye view. Another disadvantage of a field-sequential CRT stereoscopic display, or any stereoscopic display that requires the use of polarization to filter the unwanted eye view, is the reduction of luminance. A third disadvantage of a field-sequential stereoscopic display, or any display that presents images at a constant focal point while steroscopic depth is implemented through the manipulation of disparity, is the decoupling of accommodation and vergence. The problems of ghost images and the discordance between accommodation and vergence will be discussed further. Although field-sequential stereoscopic color displays suffer from these disadvantages, they are currently the most practical and viable displays for many applications.

4.2 The Advantages of Using Stereoscopic Color Displays

The role of stereoscopic depth in the perception of depth and movement is an important research question. McKee, Levi, and Bowne [MCKE90] have suggested that stereopsis is merely useful for fine hand control. Stereopsis can also break camouflage when a stationary observer has enough time to examine surroundings. Motion parallax can guide large body movements and break camouflage when an observer is moving. Nakayama, Shimojo, and Silverman [NAKA89] showed that stereoscopic depth plays an important role in delineating and linking parts of an object that is partially occluded by another object. Moreover, human observers have been found to divide their attention between different viewer-centered depth planes, as their visual performance depends upon the observer-distractor distance as well as on the target-distractor distance [ANDE90]. Because of the ability to divide attention between depth planes, the visual search for a target on one depth plane was not influenced by the distractors on another depth plane [NAKA86].

Given that stereoscopic depth can function as an effective information-coding dimension, a reliably generated depth dimension on an electronic visual display could provide performance benefits for many display tasks. By simulating three-dimensional (3D) objects or visual scenes, stereoscopic displays should provide 3D information with a higher degree of fidelity than either standard 2D displays or 2D displays supplemented with perspective graphics (2.5D). The addition of binocular disparity as a "true" or unambiguous depth cue should allow the portrayal of 3D relationships in more integrated and veridical display formats. The user could then use these advanced formats to directly visualize the spatial relationships among objects within the viewing volume and resolve any ambiguities in monocular visual cues without complex cognitive processing.

The benefits of utilizing stereoscopic depth in representing 3D spatial information have been demonstrated with various stereoscopic systems. For example, Pepper, Smith, and Cole [PEPP81] showed that stereoscopic television displays were superior to monoscopic TV displays for controlling an undersea manipulator, and that the advantage of the stereoscopic TV display became more pronounced as scene complexity and object ambiguity increased. Kim, Ellis, Tyler, Hannaford, and Stark [KIM87] demonstrated that viewing stereo pairs of two objects in a three-axis tracking task with a stereoscope led to better performance than tracking the two objects in a perspective format when monocular information was ambiguous.

Studies that use electronic field-sequential stereoscopic displays also reveal performance improvements in distinguishing the front and back surfaces of a wireframe globe image [WAY88], in positioning a 3D cursor [BEAT87], and in counting the number of target aircraft in a designated sector of an air-to-air tactical situation display [ZENY88]. Using a perspective display format at three different viewing orientations with both monoscopic and stereoscopic presentation, we [YEH92] found that stereoscopic depth facilitated spatial judgments of relative depth and altitude between two targets embedded within a displayed 3D volume. The only situation for which we did not find performance benefits from stereopsis was where subjects made judgments along the z-axis with the center of the perspective projection high above the ground, from a bird's-eye view. In this case, the task of deciding which of the two targets was in front of the other on the distance axis could be made easily by comparing the relative positions of the two objects on the ground grid. In other situations, we estimated that improvements in judgment time from the addition of binocular disparities ranged from 21.9 to 41 percent. Further, incorporating stereoscopic depth improved performance to a greater extent when monocular cues were ineffective or ambiguous [YEH92].

Although the benefits of using a stereoscopic display technology are generally acknowledged and are supported by performance data, many visual and perceptual issues related to such devices remain unresolved. Moreover, the empirical database derived from applied vision or human factors research that currently exists for developing cogent application guidelines for a new display technology is fragmented and incomplete.

4.3 Visual and Perceptual Issues in Using Stereoscopic Displays

In this section, many visual and perceptual issues related to the use of stereoscopic displays are reviewed and discussed. The understanding of stereoscopic processing will ultimately prescribe how information should be coded and presented on stereoscopic displays. Further, the understanding of stereoscopic processing may suggest conditions under which the addition of binocular disparities may be beneficial to human performance.

Stereopsis

Stereoscopic depth is induced by a horizontal difference in the position of corresponding monocular retinal images (see [BOFF88], [PATT92a], and [ARDI86] for detailed definitions and reviews). As shown in Figure 4.1, lateral retinal image disparity is defined by the difference between the fixation vergence angle and the vergence angle associated with an object in space. When an observer fixates at a certain distance, objects on a surface in space would produce corresponding monocular retinal images with zero lateral retinal image disparity. Thus, objects on this surface, defined as the horopter, are perceived as locating at the same distance as the fixation point. The shape of a horopter depends on the viewing distance. Objects in front of the horopter produce crossed retinal disparity, and objects behind the horopter result in uncrossed retinal disparity. As the viewer shifts

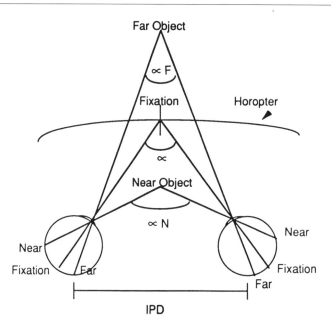

Figure 4.1
Geometric definition of lateral image disparity.

the fixation point, the retinal disparity changes accordingly. Therefore, the relation between lateral retinal disparity and depth varies as a function of viewing distance, and a given disparity value may induce various depth sensations depending upon where the fixation plane is.

As Patterson and Martin [PATT92a] have rightfully pointed out, stereoscopic depth based on the horopter and corresponding points is the only relevant definition for the visual system because disparity is based on retinal coordinates. However, the horopter and corresponding retinal points cannot be defined a priori, and dynamic measurements of horopter and vergence angle may not be practical in many display applications. Further, it has been shown that relative rather than absolute disparity is more important in stereopsis. Thus, the baseline of zero disparity is defined as the face of the display screen in many applied settings where the relative disparity between objects is of interest [YEH90]. As shown in Figure 4.2, the magnitude of depth could be computed from the geometry based on the viewing distance, separation between the two eye views, and the interpupillary distance (IPD) [CORM85]. At a given viewing distance, stereoscopic depth is considered to be most useful at portraying small depth intervals in the range of 5–10″ of arc. Thus, stereopsis could provide form and structure information in camouflaged surfaces. Binocular disparity and luminance have also been shown to provide information for the perception of motion in depth [CHAN90]. Therefore, if used properly, stereoscopic depth should be an effective coding dimension for many information display applications where the understanding of three-dimensional spatial relations or structure is crucial.

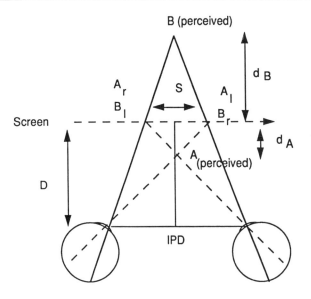

S: Separation between the two eye-views   D: Viewing distance to the screen

A_r : Right eye view of A A_l : Left eye view of A

B_r : Right eye view of B B_l : Left eye view of B

For crossed disparity: $S/d = IPD/(D-d)$

    Perceived depth, $d_A = (S \times D) / (IPD + S)$

For uncrossed disparity: $S/d = IPD/(D+d)$

    Perceived depth, $d_B = (S \times D) / (IPD - S)$

Figure 4.2
Perceived depth
computed from
the geometry.

Accommodation and Vergence

In real-world viewing, vergence changes (the rotation of the eyes inward or outward) are linked to accommodative responses (changes of the lenses of the eyes) as an observer changes the fixation distance. A mismatched relation between accommodation and vergence may cause discomfort, image blurring, or double vision [BOFF88]. Imbalanced adaptation of accommodation and vergence could produce disturbances in binocular vision [SCHO88]. In implementations that present images on a computer screen or in which two images are projected to a mirror and then reflected into one stereoscopic image, stereoscopic depth is produced by the direct manipulation of retinal disparity at a constant image focal plane (i.e., the screen or the background verged from the two eye views). Thus, the accommodation and vergence mechanisms are decoupled. The extent and the effects of this mismatch between accommodation and vergence in the use of electronic stereoscopic color displays have yet to be empirically addressed.

Using random-dot stereogram stimuli, Shea [SHEA90] found that experienced subjects showed an increase in accommodation as the disparity increased, but the magnitude lagged behind the accommodative changes to physical targets at the equivalent perceived depth. Using both a reading/search task and a vernier tracking task on an electronic stereoscopic display, our preliminary data (unpublished) showed that observers tend to accommodate to the screen distance while subjective vergence is affected by the manipulation of disparity. If it is true that accommodation tends to focus at the screen distance while vergence follows the disparity, visual fatigue could result from prolonged viewing due to the forced vergence effort [TYRE90]. In order to reduce potential visual discomfort or fatigue, binocular disparities should be manipulated near the focal image plane so that the imposed vergence demand is less straining.

Extinction Ratios and Interocular Cross Talk

Visual fatigue may also result from viewing stereoscopic images when there is some interocular cross talk between the two eye views. Because of the speed in switching on and off of the liquid-crystal materials in the shutter and the phosphor decay rate of the color monitor, ghost images may be perceptible when images are presented in green, in white, or toward the bottom of the field-sequential electronic stereoscopic monitor. Residual images to the previous eye view are visible as the raster lines sweep to the end and switch to present the imagery to the other eye view. The leakage between the two eye views can be measured by the extinction ratio, which is defined as the ratio of the luminance of the correct images over the luminance of the ghost images seen through the other eye view. The lower the extinction ratio, the higher the interocular cross talk.

Interocular cross talk could be considered noise as information from the monocular channels is summed and integrated. Using a neutral gray background to mask perceptible ghost images in various experiments, we [YEH90] found that interocular cross talk still affected fusion response when the vergence mechanism was involved. Interocular cross talk did not affect depth matching given the viewing contexts used. However, interocular cross talk strongly affected subjective ratings of image quality and visual comfort even though most of the ghost images were masked. Whether interocular cross talk would affect visual performance for other types of tasks remains undetermined. It is our experience that ghost images are highly visible and could affect visual performance when stereoscopic images are drawn in saturated green or blue over a dark background. Furthermore, asymmetrical interocular cross talk (i.e., more leakage from one eye view compared to the other) was detected in the use of some passive polarized viewing glasses.

The subjective rating data from our early studies [YEH90] suggest that interocular cross talk on field-sequential stereoscopic displays is an important design consideration. This technology-dependent problem may affect an observer's ability to fuse the two separate eye views into a single, integrated image in some viewing contexts. It also opens to question the accuracy and reliability of depth cues produced by such display systems. Additional concerns involve the perceived image quality of field-sequential stereoscopic displays and the potential for display-induced visual stress over prolonged viewing periods, concerns that can influence the ultimate performance and user acceptance of the technology. Thus, interocular cross talk should be eliminated as much as possible to minimize potential fusion difficulty and visual discomfort. Before the manufacturers replace slow phosphors such as P22 green, designers should consider the use of a neutral gray background to mask the ghost images.

Fusion Limits

The most fundamental issue in using any of the stereoscopic color displays is the selection of disparity in coding images so that viewers can fuse the two eye views easily without perceiving double images. Fusion limits have been investigated under various names, such as Panum's fusion area [MITC66a] or diplopia threshold. Diplopia threshold is usually defined as the threshold between single and double vision [DUWA81]. It is well recognized in basic visual sciences that fusion limits vary as a function of many parameters [ARDI86], including image size (spatial frequency), temporal frequency, eccentricity, illumination, involvement of the vergence mechanism, and practice. However, fusion limits appear to remain relatively constant over a broad photopic luminance and contrast range [MITC66b].

Using stimulus size and luminance contrast typical of the parameters used in graphical presentations on electronic information displays, we [YEH90] found that the mean fusion limits for ten observers were approximately 27' of arc for crossed disparity and 24' of arc for uncrossed disparity when vergence responses were eliminated in a brief viewing duration. With a 2 s stimulus duration, vergence responses enabled much larger displacements to be brought within the range of fusion, resulting in mean diplopia thresholds of about 4.93° for crossed disparity and 1.57° for uncrossed disparity. Interocular cross talk was found to affect fusion limits in the 2 s stimulus duration.

Although observers were able to fuse large horizontal disparities with large images for longer viewing intervals, the use of an extreme disparity range on a display would result in diplopia for portions of the image falling outside the limits of fusion for short-duration fixations, and would also require frequent, large changes in vergence during scanning of the display. Further, an extreme range of large disparity would not be characteristic of the types of depth coding used in electronic visual displays of relatively small angular extent. On the other hand, an extreme range of small disparity (e.g., 6' of arc) would limit the utility of stereoscopic displays even though such selection would guarantee the fusing of images for symbols of various sizes.

With complex images composed of symbols of various sizes, we recommend that designers select disparity of zero up to 20' of arc crossed or uncrossed. With these conservative yet realistic fusion limits, viewers should be able to effectively use binocular disparities in perceiving complex images or in scanning for display elements across the screen. Using disparities within this range, we [YEH92] found that stereopsis enhanced spatial judgments when monocular depth cues were ambiguous. When disparity is modulated in time at a fast rate such as 5 Hz, designers may want to reduce the disparity range based on the data from basic vision research [SCHO81].

Binocular Rivalry and Suppression

In the fusion process, the visual mechanism needs to solve the correspondence problem. When the disparity is very large, objects appear double, with two monocular images striking corresponding regions of the retinas. Binocular rivalry arises when very different half-images are presented to the corresponding regions of the two retinas and the visual mechanism would suppress one eye view. Although we rarely notice the conflict, binocular rivalry and suppression coexist with stereopsis in natural binocular viewing conditions. Whether fusion preempts suppression or suppression produces fusion has been of theoretical interest [BLAK89].

Empirical evidence shows that binocular rivalry and suppression arise when the two eye views (or half-images) are very different in attributes such as orientations of contour, size, spatial frequency, brightness, and hue. When binocular rivalry occurs, viewers perceive alternate eye views or a combined percept composed of dominant portions of each eye view. The implications of binocular rivalry and suppression for information coding are straightforward. Designers should avoid presenting large disparity or very different attributes to the two eye views. Any large interocular differences in visual attributes may cause binocular rivalry and suppression. Moderate differences in spatial frequency or orientation may also produce unwanted effects. Halpern, Patterson, and Blake [HALP87] showed that 20–25 percent of differences in spatial frequency between the two eye views could result in positional disparity, which in turn produces the perception of surface tilt.

Chromostereopsis

The perception of depth can result from retinal disparity as well as from many other cues, such as linear perspective, occlusion, size, motion parallax, and texture gradient. Depth perception can also result simply from viewing colored objects. Because our primary line of sight is 5° nasal to the optical axis of the eyes, the fovea does not usually lie on the optic axis of the eye [WEST86]. Light of different wavelengths from the fixation point would enter the eye deviated in proportion to the refractive index of the eye media for the wavelengths. With symmetrical angular displacements of the optical axes of the two eyes, colored images on the two retinas would have disparity and therefore produce depth perception.

Chromostereopsis, the difference in apparent depth of coplanar colored stimuli ([OWEN75], [VOS66]), is problematic because of individual differences in depth localization of colored targets. Some viewers perceive targets of short wavelength (blue) as near and targets of long wavelength (red) further away, and other viewers may perceive the opposite depth relationship. Further, the magnitude and the direction of this depth effect may vary with changes in illumination [YE91]. As for the individual differences, it has been shown that the interocular difference in monocular transverse chromatic aberration can account for chromostereopsis with small pupils [YE91]. To avoid chromostereopsis, Murch's [MURC82] data on accommodation and vergence to CRT display colors suggest that designers should use secondary colors rather than saturated primary colors in coding stereoscopic images. If a designer attempts to use colors to code depth, empirical investigations should be conducted to map the relation between depth perception and color manipulation prior to the selection of color coding schemes. Depending on the spectral composition in the stimuli and the illumination, observers may perceive the reverse depth effect from what is intended. Because of this potential problem and individual variability, it is not recommended that colors be used as the only medium to code depth.

Stereoacuity

Stereoacuity is the smallest resolvable difference in depth that viewers could detect between two targets. Stereoacuity, or stereothreshold, is an exponential function of a target's distance from the fixation depth plane [BLAK70]. Around the fixation plane, stereoacuity is about 4–5″ of arc near the fovea [MCKE83]. Lower stereoscopic acuity, around 1–2″, has also been reported [BERR50]. Stereoacuity increased exponentially as the target was moved further in depth from the fixation plane. Using a relative depth judgment task with zero disparity as reference, we [YEH90] found that the relation between the accuracy of relative depth judgment and the magnitude of disparity was well approximated by an exponential function for both crossed and uncrossed directions, with a greater variability and a shallower slope for uncrossed disparity.

The exponential function may actually consist of two segments. Badcock [BADC85] found that stereothreshold increased at the rate of 3″ of arc per 1′ arc-of-disparity increments up to 20–40′ of arc. Stereothreshold remained relatively constant beyond this disparity value. Data from McKee [MCKE90] also demonstrated two segments of stereothreshold as a function of test disparity. Up to 10–20′ of arc of disparity, stereothreshold was approximately 5 to 10 percent. In other words, subjects could discriminate a test target 1.0–2.0′ of arc from another object that was 20′ of arc in disparity from the fixation plane. Stereothreshold remained relatively constant for disparities of 20′ to 60′. Given that we recommend the use of disparity up to 20′, the least sensitive stereothreshold would be approximately 2′ of arc.

Thus, viewers should be able to discriminate various objects on the fovea along the depth dimension when the disparity difference between objects is around 2′ of arc. Larger disparity separation between objects should further facilitate depth discrimination near the fovea.

In addition to the disparity from the fixation plane, many other factors influence stereoacuity. These factors include the presence of adjacent objects near the test stimuli, the spatial separation between the fixation point and the test stimulus, the luminance of the test stimuli and visual surround, and stimulus orientation [ARDI86]. Further, factors such as test stimulus configuration ([MCKE83], [MITC84]), practice [FEND83], spatial frequency content [BADC85], eccentricity, temporal modulation, and the spatial separation of test targets [FEND83] all significantly affect stereoacuity. Larger disparity separation between objects may be necessary to produce accurate depth discrimination for large targets or targets presented in the periphery. Fine stereoacuity of a few seconds of arc can be achieved only with small or thin objects near the fovea around the fixation plane.

If few isolated objects are used in an application, the designers could analyze the viewing context and consult the literature to approximate the optimal disparity separation between objects at various eccentricity positions. If many elements are used in the application, the designers should empirically investigate the optimal disparity separation for information coding until a large human performance database is developed that can be used to map the function between stereothreshold and various visual parameters for different task environments.

Crossed and Uncrossed Disparity

It has been suggested that crossed disparity (perception in front of the fixation point) and uncrossed disparity (perception behind the fixation point) are processed by separate mechanisms and that the encoding of uncrossed disparity is less accurate [MUST85]. Neurophysiological evidence suggests that there are six types of cells: tuned zero, tuned inhibitory, near, far, tuned near, and tuned far [POGG88]. Whether these cells are responsible for disparity processing remains unresolved. Freeman and Ohzawa [FREE90] have suggested that binocular disparity is encoded by cells in terms of phase disparity at each spatial frequency scale. Human performance data also show that the difference in processing crossed and uncrossed disparity is still an open research question.

Patterson and his colleagues [PATT92b] found that depth effect in the uncrossed direction was frequently less than predicted from the geometry unless large stimuli were presented with a long duration. In conjunction with the viewer-centered allocation of attention in three-dimensional space, designers may want to utilize crossed disparity to code most of the elements. However, Parrish and Williams [PARR90a] reported that depth perception in the uncrossed direction was greater than predicted by the geometry. McKee and her colleagues [MCKE90] did not find any direction difference in stereothreshold. Yang and Blake [YANG91] showed that two spatial-frequency channels (3 cycles/degree and 5 cycles/degree) were observed for both crossed and uncrossed disparities from coarse to fine. Also, Stevenson and colleagues [STEV92] did not find a direction difference in the detection of interocular correlation in random-dot stereograms.

We [YEH90] found that depth matching among elements in a row of a matrix was less accurate with uncrossed disparity. When subjects were asked to judge relative distance from the screen, judgments with uncrossed disparity were more variable, but there was no direction difference. It appears that depth judgments depend on the vergence angle imposed by the disparity relative to the fixation point. It is plausible that depth judgments for uncrossed disparity could be as accurate as judgments for crossed disparity within a certain range of vergence demand. It should be noted that the direction of disparity is based on the fixation point (or horopter). An uncrossed disparity may become crossed as the viewer shifts the fixation point behind the uncrossed disparity. It is evident that we need more empirical research to disclose the complex relation among disparity, the direction of disparity, vergence, size, stimulus duration, and the visual task.

Disparity and Luminance Contrast in Depth Perception

In addition to the direction of disparity, vergence, stimulus size, and fixation plane, other visual parameters may also affect the effectiveness of binocular disparity in providing the depth percept. Stevens and his colleagues ([BROO89a], [STEV88]) found that monocular depth cues in linear perspective dominate in the presence of conflicting binocular disparity in spatial judgments of planar surfaces. Based on the findings, they proposed that depth is reconstructed from boundary contrast features defined by disparity discontinuities analogous to brightness contrast in edge detection. Consistent with this finding, the disparity of the stimulus edges appears to govern the solution of the correspondence problem [MCKE88]. As the visual system solves the correspondence problem, objects in space will either fall on the horopter or produce lateral disparity on the retinas. It appears that contrast in both disparity and luminance is important in stereopsis. Bulthoff and Mallot [BULT87] showed that depth perception is accurate when stereoscopic depth can be derived from the relative locations of intensity edges in stereo images. When sharp luminance contrast is reduced around the edges of a smooth-shaded sphere, stereoscopic depth could still be derived from the intensities but with lower accuracy. Legge and Gu [LEGG89] showed that disparity sensitivity was correlated with luminance contrast sensitivity across a range of spatial frequencies. Reversing the luminance contrast of a random-dot stereogram could reverse its apparent depth [BOWN90].

However, more empirical research is necessary to identify the relationship between luminance contrast and stereopsis in depth perception given that difference in luminance contrast does not affect fusion. Furthermore, Yang and Blake [YANG91] showed that there are only two luminance spatial-frequency mechanisms in stereopsis.

Interaction in Disparity Processing

At the local level, disparity between two components sharing similar visual attributes could be pooled and averaged. Mansfield and Parker [MANS90] showed disparity averaging in viewing a random-dot stereogram between two components sharing identical orientation features. Local disparity processing could also be influenced by global disparity interpretation. Stereopsis on coarse (low-spatial-frequency) scales constrained processing on fine (high-spatial-frequency) scales. Fusion limits to a grating pattern of high spatial frequency at 10–12 cycles/degree were found to be constrained by a mask of frequency two octaves below, at 2.5–3 cycles/degree. Masks with spatial frequencies higher than the test grating pattern did not affect the fusion limits to the pattern [WILS91]. Furthermore, differences in color and temporal frequency did not reduce the constraints imposed by the mask on the test grating pattern [ROHA90].

Both interaction at the local level and the imposition of constraints from the global to the local level could be evident when many elements are displayed on the screen. It has been suggested that global, coarse disparity may be used in depth segmentation at an early level for which a set of matches at the local level is chosen to approximate a plane [MITC88]. Depth matching data also show that both global and local factors influence depth perception ([MITC84], [FAHL88], [MITC87]). Consistent with the findings in basic vision research, we [YEH90] observed that disparity context influenced depth matching of elements in an array displayed on a field-sequential stereoscopic display.

Schor [SCHO91] has identified three types of global interaction: disparity crowding, disparity gradients, and disparity interpolation. Disparity crowding occurs when targets are separated in xy space by less than $10'$ of arc. Disparity gradient operates in the separation of 10–$15'$ of arc; over this range disparity processing depends on the spatial separation and the difference in disparity between objects. Finally, disparity interpolation may be observed under conditions where disparity differences between neighboring regions are too gradual to be detected (consistent with the observation that discontinuities in disparity are important for depth perception), where stereo patterns are presented too briefly, or where ambiguous disparity solutions are presented in a region. Depth interpolation may also interact with spatial frequency [WURG89]. The extent of interaction also depends on the retinal eccentricity, or distance from the fixation point. Manipulating the eccentricity of a center point and the spatial separation between the point and a hexagonal array, Westheimer and Truong [WEST89] found that stereoacuity was influenced by both manipulations. Their results showed that the spatial separation needed for best stereoacuity (a few seconds of arc) was about 20′ when the center point was presented at the fovea and about 110′ of arc when the center point was presented at 6–9° in the periphery. The worst-case stereothreshold for judging whether the center point was in front of or behind the surround hexagonal array was about 110″ of arc for a target point presented at 9° and spatially separated 10′ from the array. Therefore, a disparity separation of 2′ should allow observers to accurately distinguish the depth of a point from the surround at positions up to 9° in the periphery.

Size Perception

According to the size-distance invariance hypothesis, the ratio of perceived size to perceived distance is constant for a given visual angle (see [EPST77] for a thorough discussion on constancy in visual perception and [SEDG86] for a review of space perception). Given a constant visual angle, the perceived size of an afterimage is proportional to the perceived distance to the surface on which the afterimage is projected (Figure 4.3). This relation is known as Emmert's law. Since objects of identical physical size (same retinal visual angle) would be perceived as located at various

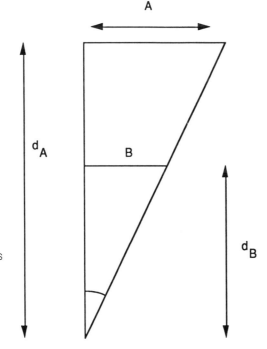

Figure 4.3
Emmert's law of size constancy. Given the same retinal angle, B is perceived as smaller than A because B is perceived as closer than A.

distances according to the disparity on a stereoscopic display, size perception should also follow Emmert's law. Using a random-dot stereogram, Shea and her colleagues [SHEA82] found that convergence angle increased with increases in crossed disparity from 20′ of arc to 3.6°. Further, size judgments for a random-dot stereogram of constant retinal size decreased with increases in crossed disparity, and scanning eye movements followed the perceived size. Asking subjects to estimate the distance and size of a square with a standard at zero disparity, we found that perceived size also followed Emmert's law, although there were individual differences in the slope linking the estimated size to the disparity manipulation (unpublished data). In general, a test square of constant physical size was perceived as smaller when the test object had crossed disparity relative to the screen and was perceived as larger when the test object had uncrossed disparity.

As demonstrated in the literature, the size-distance invariance hypothesis does not hold in many situations (see [SEDG86] for a review). The relation between disparity and size is becoming more intriguing as vision scientists disclose the underlying mechanisms relating disparity and spatial-frequency (size) channels. On one hand, it was found that there are only two independent spatial-frequency mechanisms for stereopsis [YANG91]. On the other hand, it appears that there is a disparity-size correlation in the fusion process. Data from [SCHO81] show that fusion limits are smaller at high spatial frequency (smaller size) and larger at low spatial frequency (larger size). However, fusion limits may depend more on the low-spatial-frequency component in an image. It has been found that blue cones can support stereoscopic vision but do not provide input to high-spatial-frequency mechanisms [BLAK87]. Further, a mask of spatial frequency two octaves below would constrain the fusion limits to a pattern of high spatial frequency [WILS91]. In a model of stereo processing, a set of six crossed and six uncrossed disparity sensors were used. Disparity sensors combine information from two monocular spatial filters that share similar spatial-frequency characteristics [BOWN90]. The exact nature of the visual system's encoding of binocular disparity and spatial frequency is an interesting research area that may reveal the mechanisms underlying depth and size perception in three-dimensional visual space.

When size constancy is of critical importance in the performance of a task, size scaling should be empirically determined for each user. When we presented test squares in sizes compensated by the inverted Emmert's law (i.e., larger size for crossed disparity and smaller size for uncrossed disparity according to the geometry of stereoscopic depth), we predicted that subjects would perceive the size of a test square located at various depths to be the same as the size of the standard square with zero disparity. However, we found individual differences in size perception (unpublished data). Some subjects perceived the test square in its physical size (larger square for crossed disparity and smaller for uncrossed disparity), and other subjects perceived the test square to be the same size as the reference. Further, the individual differences in size perception could not be attributed to the differences in distance estimates. Therefore, individual differences in size perception may be due to each subject's interpretation of the task. Some subjects may use the retinal size, and others may scale distance in estimating the size.

Dynamic Stereopsis

As relative contrast is important in processing static disparity, perception of motion in depth also requires information about relative motion [REGA86]. Regan and his colleagues ([REGA73], [REGA86], [HONG89]) have demonstrated that the human visual system is specifically sensitive to the ratio between the velocities of right and left retinal images that are moving in depth. Moveover, receding and approaching motions in depth are processed by different mechanisms. Dynamic depth perception is reduced at temporal frequencies higher than 1–2 Hz. More interestingly, some subjects with normal sight were blind to changing disparity. How this stereo motion blindness is linked to retinal inhomogeneity is an intriguing question, since Kelly [KELL89] showed that retinal inhomogeneity may play an important role in processing information in egocentric motion along the line of sight. Consistent with Regan's proposal, Kelly showed that there may be independent forward and backward channels for egocentric motion, with the forward channel better developed. Furthermore, Kelly showed that low spatial frequencies (0.05–0.5 cycle/degree) are involved in forward-motion detection over the temporal modulation of 1–8 Hz.

Empirical findings suggest that temporal modulation in stereopsis should be around 1 Hz and no faster than 2 Hz. Manipulating luminance contrast in dynamic random-dot stereograms, Sato, Nishimura, and Ohzawa [SATO90] showed that the latency of visual evoked potentials to the onset and offset of disparity transients was about 200 ms, followed by a slow, sustained component when the luminance contrast is high. Therefore, dynamic changes in disparity should be accurately processed at the rate of 1–2 Hz. Furthermore, it appears that luminance also plays an important role in the perception of stereoscopic motion ([SATO90], [CHAN90]). Thus, luminance should be comodulated with disparity in the presentation of dynamic information on stereoscopic displays. Variables other than luminance may also influence depth perception. These variables include dynamic occlusion and relative motion [ONO88], interocular sequence [SHIM88], and eye of origin [NAKA89]. From the coding perspective, these variables are not a great concern as long as the dynamic changes mimic the changes in the natural scene.

Interaction Between Stereopsis and Other Visual Processing

Early observations from anatomical and physiological studies suggested the existence of independent pathways for processing binocular disparity separate from the processing of color and form [LIVI88]. Based on the evidence, Tyler [TYLE90b] suggested three mechanisms: the colorblind Magno channel processes stereo motion and coarse disparity, the colorblind Parvo-interblob channel encodes high spatial frequency and static disparity, and the Parvo-blob channel processes chromatic information at low spatial frequencies.

However, recent studies from both neuroscience and psychophysics suggest that the pathways for encoding binocular disparity may not be as independent as the early evidence suggested. It has been shown that color provides weak signals to stereopsis [SATO90]. Using an ideal-observer analysis, Jordan and colleagues [JORD90] found that chromatic cues were used with either equal or higher efficiency than luminance cues in solving the correspondence problem (i.e., fusion) in an ambiguous stereogram. Using both lesion and single-cell recording methods on rhesus monkeys, Schiller and his colleagues ([SCHI90], [LOGO90]) have shown that the color-opponent channel is essential for processing color, texture, high spatial frequency, and fine stereopsis, whereas the achromatic broadband channel is important for processing fast flicker and motion. How such neural processing would affect human information processing of these visual cues at the higher level is unknown at the present time.

At the perception and performance level, two paradigms are generally used to study the interaction between stereoscopic depth and other visual cues. The first paradigm presents conflicting information to investigate which visual cue dominates over other cues in resolving the final percept. The second paradigm applies the factorial design to study whether stereopsis interacts with other visual processing. Results from the first paradigm have generally shown that binocular disparity, occlusion, and motion are dominant cues for depth perception. Depending on the type of occlusion, surface, and structure that velocity represents, stereopsis could be dominated by occlusion, motion, and perspective. Further, factors such as reliability and consistency among estimates of depth may influence how multiple cues are integrated in providing the final percept [LAND91]. Results from the second paradigm have usually shown an additivity effect, in which all depth cues independently contribute to depth perception, with one exception: the addition of motion could interact with other cues in affecting depth perception [WICK90].

Disparity Curvature and Object Recognition

In the study of stereopsis, isolated binocular features are often used in the stimulus configuration to investigate how binocular disparity could provide relative depths of objects in the visual scene. However, the investigations of isolated disparity may not be relevant to understanding how the visual system perceives disparity modulation in space. Further, research has begun to reveal the importance of stereopsis in the perception of visual surfaces. The investigation of disparity surface and curvature can be traced to the findings that there may be independent stereoscopic mechanisms in the visual system that are sensitive to different spatial frequencies of disparity modulation in space [SCHU79]. By varying the modulation of disparity or motion parallax, Rogers and Graham [ROGE82] showed that both motion parallax and stereopsis are involved in interpreting the structure of three-dimensional surfaces with maximum sensitivity when the spatial frequency of the corrugations is about 0.2–0.4 cycle/degree. Further, Rogers and Cagenello [ROGE89] empirically demonstrated that disparity curvature is independent of the viewing distance, and subjects could easily discriminate and match disparity curvature. Being invariant over the viewing distance, disparity curvature is advantageous over isolated binocular disparity, which can shift from one direction to another if the viewing distance changes. Using the conflicting paradigm, Stevens [STEV90] showed that monocular interpretation of a surface tends to dominate at locations where the disparity pattern indicates planarity. Stereopsis dominates at locations where the disparity pattern suggests curvature and the monocular surface indicates planarity.

Evidence clearly suggests that the second spatial derivatives of disparity gradients (i.e., disparity curvature) may be a primary cue for both depth and shape perception. When the second derivative is zero (no curvature), depth matching of two target lines was influenced by the disparity gradients in the background ([MITC84], [BROO89b]). Gillam and her colleagues [GILL88] also showed that subjects had difficulty in judging the surface slant when there was no discontinuity in the gradients. Consistent with the analogy of brightness contrast, judgment of surface slant was much easier when there were discontinuities at a surface boundary. More empirical research is needed to disclose the underlying mechanisms for integrating isolated disparity into disparity gradients based on which disparity curvature is derived. Alternatively, the visual system could begin with the disparity discontinuities and gradients, and then reconstruct isolated depth [BROO89b].

Individual Differences

In addition to the visual and perceptual issues, the effectiveness of using stereoscopic depth in image representation depends on user characteristics. Stereoscopic displays are appropriate only for viewers who have normal stereo vision, just as normal color vision is required for the viewer to take advantage of colors in display applications. It should be noted that large individual differences in stereopsis are found in the population. For example, in a survey of 150 members of the student community at the Massachusetts Institute of Technology, 4 percent were unable to use the depth cue offered by disparity, and 10 percent experienced great difficulty in perceiving depth in a random-dot stereogram [RICH70]. However, Patterson [PATT84] suggested that the percentage of stereoblind observers is much smaller (one out of ninety-eight subjects) when stereoscopic images were viewed for an unlimited viewing duration.

Using observers with stereoacuity of 27″ of arc or better, we [YEH90] found large individual differences in the subjects' ability to converge and diverge their eyes to bring a large disparity into fusion. When a stereoscopic image was presented for a brief duration within a 200 ms interval, individual differences in fusion limits were much smaller and may be attributable to the leniency in the criterion chosen for judging the singleness of vision. Within-subject variability was relatively small, and there were no significant changes of fusion or depth judgments over a period of 1.5 hours given that the disparity range in coding was within the fusion limits determined from the 200 ms interval (27′ of arc crossed to 24′ of arc uncrossed).

Since stereoscopic processing is largely a visual operation, individual differences are to be expected since observers differ in their ability to make accommodative/vergence changes, in their dark vergence focus, in their interpupillary distance, in their tolerance of mismatch between accommodation and vergence, and in their sensitivity to various spatial frequencies. How such individual differences affect stereoscopic information processing is unclear as we do not have a large database from which to estimate the effect. The presence of monocular contour could facilitate the eye movement necessary for perceiving stereoscopic depth. Screening subjects on deficiency in the perception of stereo motion may be important for some applications. Tittle, Rouse, and Braunstein [TITT88] showed that subjects who were classified as deficient in static stereo vision could use stereoscopic information in displays with disparity continuously updated. Given that subjects could use changing disparity even if they are deficient in static stereo vision, stereo motion blindness may be more crucial than stereoblindness to static disparity for applications in which disparity is continuously updated. For these applications, a clinical test designed by D. Regan is recommended. Hong and Regan [HONG89] showed that some observers may be deficient in perceiving motion in depth along one direction.

4.4 Summary of Coding Recommendations

As is apparent from this short review of the visual and perceptual issues in using stereoscopic color displays, the processing of binocular disparity is complex and intriguing. Many visual parameters as well as the task context affect how disparity is processed in viewing a stereoscopic image. Therefore, the following recommendations should be taken with caution and updated as more empirical data are compiled. Designers should also consult the empirical database in the literature to specify coding strategies.

At the present time, it appears that designers could use a disparity range from zero up to 20′ of arc in both crossed and uncrossed directions. Use of a greater range of crossed disparity (e.g., up to 30–40′) may be acceptable, as some evidence suggests that depth perception is more veridical with crossed disparity and observers may allocate attention in a viewer-centered zone. If large disparities are necessary for the application, large disparity should be coded in patterns of low spatial frequency of luminance (e.g., boundary) and small disparity should be used for high-spatial-frequency of luminance (e.g., fine details). This coding scheme is consistent with coarse-to-fine disparity processing. Empirical testing of fusion limits and user acceptance at the early stage may be required to ensure the most effective use of stereoscopic displays.

Within the fusion limits, saturated primary colors should be avoided, as these colors may evoke depth that is inconsistent with stereopsis and hence eliminate the depth effect. The temporal modulation of binocular disparity should be less than or equal to 2 Hz so that the visual mechanisms can accurately process dynamic disparity. When plausible, stereoscopic images should be spatially separated to eliminate disparity averaging, repulsion, or crowding. Designers who wish to code disparity separation according to eccentricity, spatial frequency, and disparity value should consult empirical findings in the literature to map out the coding scheme. In general, a disparity separation of approximately 2' of arc at the fovea should permit many users to correctly discriminate depth among objects locating at various depth planes. The disparity separation of 2' should also allow users to correctly distinguish a target at 9° eccentricity from surrounding objects.

The drawing of stereoscopic imagery should be executed from far to near so that uncrossed disparity will not mask the crossed disparity. Further, occlusion and interpolation in the drawn visual scene would then mimic the natural visual scene. Size should be scaled according to the disparity, and size scaling should be individually determined when size constancy is crucial for the task. Luminance should also be comodulated because luminance contrast is a depth cue (bright objects are seen as nearer), and many studies have shown that luminance is an important cue in stereoscopic processing. In all cases, interocular differences in visual attributes should be eliminated to prevent binocular rivalry and suppression.

4.5 Conclusions

Stereoscopic display systems are currently being used to provide high-fidelity visualization for a variety of tasks, and it is likely that this new display technology will find much greater application in the future. In this chapter, visual and perceptual issues in the use of stereoscopic displays were briefly reviewed. Throughout the chapter, it was shown that there are many unresolved issues that require systematic investigations across the fields of neuroscience, psychophysics, and human performance. Although the study of stereopsis has attracted a great deal of attention and interest in basic visual sciences, insufficient data derived from applied vision or human factors research exist for developing cogent application guidelines. Furthermore, many of the fundamental human performance issues in stereoscopic imaging have yet to be addressed. The realization of performance advantages is critically dependent on the type of visual task, monocular and binocular cue effectiveness, and a consistent display coding strategy based on the unique perceptual attributes of binocular visual processing. As the technology attracts greater interest in the applied community, we hope that the human factors issues will be systematically investigated and addressed. Empirical data from such investigations should provide valuable guidance to display designers, human factors specialists, and potential users of stereoscopic displays.

Acknowledgment

Much of the review and discussion in this chapter was initiated while I was working in the Human Factors Display Group at the Phoenix Technology Center, Honeywell. I would like to acknowledge the inspiration of Louis Silverstein, now at VCD Sciences, who introduced me to the world of applied vision.

5

Computing Stereoscopic Views

Larry F. Hodges

David F. McAllister

5.1 Introduction

Several methods for computing the left- and right-eye views of a stereo pair have appeared in the literature. Grotch [GROT83] and Roese and McCleary [ROES79] described stereoscopic images based on rotations. Lipscomb [LIPS79] approximated rotations with shears. Baker [BAKE87] produced the left- and right-eye views with a lateral shift along the x-axis, along with a lateral shift of the resulting images relative to each other. Penna and Patterson [PENN86] have suggested choosing two different eye positions and a single viewplane.

In this paper we analyze the different approaches to computing the left- and right-eye views of a stereo pair. Each approach is first described analytically relative to a right-handed Cartesian coordinate system with a projection plane located a distance d from the origin and parallel to the xy plane. Within this framework we derive expressions for horizontal parallax, vertical parallax, and location of the stereo window.

5.2 Terminology and Definitions

Depth Cues

In a stereoscopic image, since each eye is actually focused or accommodated on a flat image at a constant distance from the viewer, accommodation cues are inconsistent with the perceived depth. Consistent convergence cues are provided by the horizontal positioning of the left- and right-eye views of a scene. Binocular disparity cues are provided by computing the left- and right-eye views from different positions relative to the scene. Based on his work with random-dot stereograms, Julesz [JULE71] rates binocular disparity as the strongest single depth cue for determining relative depth. Cavanagh [CAVA87] also rates binocular disparity as an important cue but found that the priority of conflicting cues varied from observer to observer.

It is generally accepted that accommodation and convergence are less important depth cues than binocular disparity, but that they do influence depth perception when coupled with binocular disparity, especially for observer distances less than 2 meters ([ITTE60], [JULE71]).

Horizontal Parallax

We first review the model of the relationship of binocular disparity to depth in a stereoscopic display. Imagine a three-dimensional scene that contains a clear rectangular window in a plane parallel to a plane intersecting the two eyepoints of an observer, with the eyepoints horizontally aligned. This window is the stereo window. Part of the scene is behind the window, part of the scene is in the plane of the window, and part of the scene extends in front of the window.

First consider the two lines formed by a point P in the scene and each of the observer's eye positions. Because the eyes are horizontally separated by approximately 2.5 inches, which is the interocular distance, P will be projected to different points L and R on the stereo window, corresponding to the lines of sight of the observer's left and right eyes, respectively. The distance between the abscissas of the two points, $R - L$, is the horizontal parallax. A positive horizontal parallax corresponds to a point P_1 behind the window. A negative horizontal parallax corresponds to a point P_2 in front of the window. A point P_3 in the plane of the window has zero parallax.

The maximum possible horizontal parallax is equal to the interocular distance. Maximum positive horizontal parallax places a point infinitely far behind the window. Horizontal parallax equal to the interocular distance occurs when the point is halfway between the observer and the stereo window. Note that these properties hold regardless of the distance of the observer from the stereo window.

Image Scaling along the Viewing Axis

The projection of points onto the stereo window depends on the observer's position relative to the window. This is shown in Figure 5.1. As the observer moves further away from the stereo window, the horizontal parallax for a fixed point decreases, and as the observer moves closer, the horizontal parallax increases.

Since the observer's viewing distance is usually unspecified or unknown, the left- and right-eye views are normally computed with the assumption that the observer is located in the center of and at a fixed distance from the stereo window, and (except in head-sensing displays) the views are not dynamically changed as the observer changes head position. This assumption results in a fixed horizontal parallax for points at a particular distance from the stereo window regardless of the position of the observer. This fixed horizontal parallax causes distortion of the image

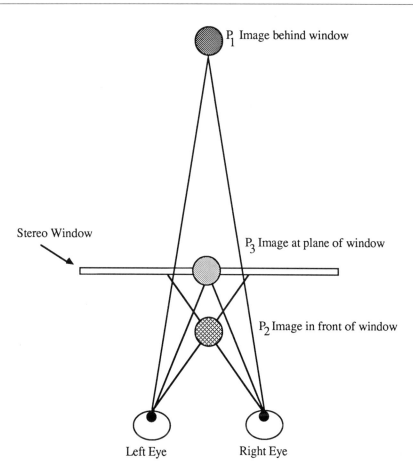

P_1 Image behind window

Stereo Window

P_3 Image at plane of window

P_2 Image in front of window

Left Eye   Right Eye

Figure 5.1
Stereo window.

along the viewing axis normal to the plane of the observer, as shown in Figure 5.2. As the observer moves away from the stereo window, the image is elongated. As the observer moves closer, the image contracts. As the observer shifts his or her head from side to side, the view of the image shifts in conjunction with the head movement. The practical implication of this is that there is an optimal position from which to view the stereoscopic image, where the front-to-back scaling of the image is in correct proportion to the up-down and left-to-right scaling [HODG92].

Vertical Parallax

Vertical parallax is the difference between the vertical coordinates of homologous points. Numerous authors insist that any vertical parallax in a scene makes fusing of the left- and right-eye views difficult and uncomfortable, especially for extended viewing sessions ([FERW82], [WAAC85]). Lipton [LIPT82] suggests that vertical parallax between homologous points should be kept to within 10′ of arc. Fender and Julesz ([FEND67], [JULE71]), using random-dot stereograms, found that experimental subjects could fuse

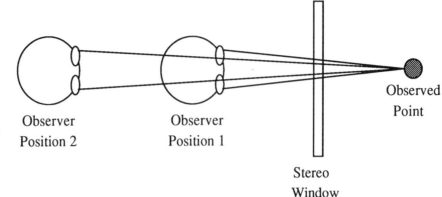

Figure 5.2
Distortion caused by head movement and fixed horizontal parallax.

Observer Position 2

Observer Position 1

Observed Point

Stereo Window

the two views with vertical parallax of 6′ of arc. Once fusion occurred, however, the two images could be separated by as much as 20′ of arc without loss of fusion.

Coordinate Systems

Analysis and results are presented in a device-independent screen coordinate system that is based on a square display screen with the origin at the center of the screen. From the origin it is unit length to the top, bottom, right, or left edge of the screen. To apply our results to a particular CRT screen, the user should scale to device coordinates. For projections we assume that the center of projection is located in the xy plane ($z = 0$). The projection plane is located a distance d from the origin and parallel to the xy plane. All calculations are done in a right-handed coordinate system.

5.3 Rotations with Perspective Projection

Motivation for This Approach

Assume that both of an observer's eyes are focused at a particular point in a three-dimensional scene. This is illustrated in Figure 5.3a, showing the top view of a rectangular box (the heavy black line marks the front of the box). The eyes are rotated inward, and the lines of sight cross at some point P, at an angle ϕ. To generate the two views we rotate the scene about the point P through an angle $\phi/2$ from right to left, for the right-eye view, and we rotate about the point P through an angle $\phi/2$ from left to right, for the left-eye view. This is shown in Figure 5.3b.

Equations for Left- and Right-Eye Projections

We will model the rotation approach as follows. Assume a right-handed coordinate system with the center of projection at (0, 0, 0) and a viewplane

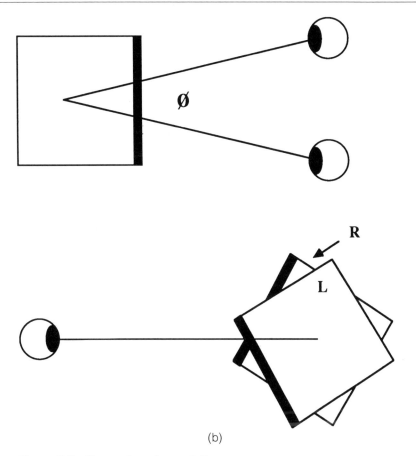

(b)

Figure 5.3 Stereo views from rotations.
(a) Right- and left-eye views of a rectangular box. (b) Views created by rotating box with respect to a single viewpoint.

parallel to the xy plane and located a distance d along the z-axis from the center of projection. We assume device-independent screen coordinates in the range $-1 \le x \le 1$ and $-1 \le y \le 1$. We do a y-rotation about a point $(0, 0, R)$ through an angle $-\phi/2$ to produce the right-eye view, and a y-rotation about the point $(0, 0, R)$ through an angle $+\phi/2$ to produce the left-eye view (Figure 5.4).

We begin by considering a point $P_0 = (x_0, y_0, z_0)$. The rotated points are

$$P_{0\text{left}} = (x_0 \cos(\phi/2) + (z_0 - R) \sin(\phi/2), y_0,$$
$$(z_0 - R) \cos(\phi/2) - x_0 \sin(\phi/2) + R)$$

and

$$P_{0\text{right}} = (x_0 \cos(\phi/2) - (z_0 - R) \sin(\phi/2), y_0,$$
$$(z_0 - R) \cos(\phi/2) + x_0 \sin(\phi/2) + R)$$

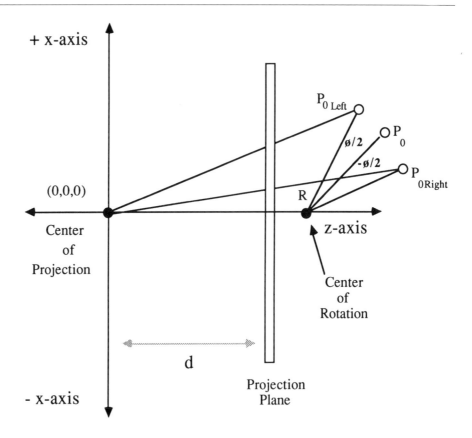

Figure 5.4
Rotation with perspective projection.

After the perspective transformation we have

$$x_{rw} = \frac{d[x_0 \cos(\phi/2) - (z_0 - R) \sin(\phi/2)]}{(z_0 - R) \cos(\phi/2) + x_0 \sin(\phi/2) + R} \tag{5.1}$$

$$y_{rw} = \frac{dy_0}{(z_0 - R) \cos(\phi/2) + x_0 \sin(\phi/2) + R} \tag{5.2}$$

and

$$x_{lw} = \frac{d[x_0 \cos(\phi/2) + (z_0 - R) \sin(\phi/2)]}{(z_0 - R) \cos(\phi/2) - x_0 \sin(\phi/2) + R} \tag{5.3}$$

$$y_{lw} = \frac{dy_0}{(z_0 - R) \cos(\phi/2) - x_0 \sin(\phi/2) + R} \tag{5.4}$$

where (x_{rw}, y_{rw}) are the coordinates of point P_0 relative to the right eye, and (x_{lw}, y_{lw}) are the coordinates of point P_0 relative to the left eye.

Vertical Parallax

We will show that rotations with perspective projection introduce vertical parallax between the left- and right-eye views. This effect has been pointed out by several others, including Baker [BAKE87] and Butts and McAllister [BUTT88a]. We derive an exact expression describing the amount of vertical parallax in terms of R, d, and ϕ. This expression allows us to compare the maximum vertical parallax to the maximum allowable parallax based on the results of Fender and Julesz [FEND67].

The vertical parallax, V, is equal to $y_{lw} - y_{rw}$ and, from Equations (5.2) and (5.4), is given by

$$
\begin{aligned}
V &= \frac{d y_0}{(z_0 - R)\ \cos(\phi/2) - x_0\ \sin(\phi/2) + R} \\
&\quad - \frac{d y_0}{(z_0 - R)\ \cos(\phi/2) + x_0\ \sin(\phi/2) + R} \\
&= \frac{2\ d x_0 y_0\ \sin(\phi/2)}{[(z_0 - R)\ \cos(\phi/2) + R]^2 - x_0^2\ \sin^2(\phi/2)}
\end{aligned}
\tag{5.5}
$$

From the numerator of this expression we note that the vertical parallax is zero when either x_0 or y_0 is zero. Furthermore, V is an increasing function of $x_0 y_0$ and a decreasing function of z_0. Since the parallax increases as the denominator tends toward zero, it follows from Equation (5.5) that large vertical parallax occurs when x_0 is approximately equal to

$$
\frac{z_0 + R[1 - \cos(\phi/2)]}{\sin(\phi/2)}
\tag{5.6}
$$

Since ϕ is always small, $\sin(\phi/2)$ is small, which means that x_0 must be large compared to z_0 for Equation (5.6) to be close to zero. Since, with device-independent screen coordinates, viewable projected points have coordinates (x_w, y_w) such that $|x_w| \leq 1$ and $|y_w| \leq 1$, and the projection is of the form $x_w = xd/z$ and $y_w = yd/z$, this case may be ignored when $d \geq 1$.

The maximum amount of vertical parallax for a point near the stereo window ($z = R$), measured as a percentage of the height of the CRT screen, is given in Table 5.1 for various values of R and d. The table assumes that $\phi = 6°$. The amount of vertical parallax is greatest at the corners of the CRT screen and diminishes toward the center. Smaller values of ϕ decrease the maximum vertical parallax, but they also decrease the horizontal parallax, which produces the stereoscopic depth effect.

In their experiments with random-dot stereograms, Fender [FEND67] and Julesz [JULE71] found that a breakdown of fusion of the right- and left-eye images occurs with vertical parallax of about 20′ of arc. If we assume that the observer is located a distance R (scaled to device coordinates) from the CRT screen, then this value, as a percentage of CRT screen height, may be calculated as $R(\tan\frac{1}{3})/2$. However, they also found that once breakdown occurs, the views do not refuse unless they are brought much closer together,

Table 5.1 Maximum amount of absolute vertical parallax with a rotational angle of 6° and $z = R$, measured relative to normalized screen coordinates for values of R and d between 1 and 10.

						R					
		1	2	3	4	5	6	7	8	9	10
	1	.1050	.0262	.0116	.0065	.0042	.0029	.0021	.0016	.0013	.0010
	2	.2099	.0524	.0233	.0131	.0084	.0058	.0043	.0033	.0026	.0021
	3	.3149	.0786	.0349	.0196	.0126	.0087	.0064	.0049	.0039	.0031
	4	.4198	.1047	.0465	.0262	.0167	.0116	.0085	.0065	.0052	.0042
d	5	.5248	.1309	.0582	.0327	.0209	.0145	.0107	.0082	.0065	.0052
	6	.6298	.1571	.0698	.0393	.0251	.0174	.0128	.0098	.0078	.0063
	7	.7347	.1833	.0814	.0458	.0293	.0204	.0150	.0114	.0090	.0073
	8	.8397	.2095	.0931	.0523	.0335	.0233	.0171	.0131	.0103	.0084
	9	.9446	.2357	.1047	.0589	.0377	.0262	.0192	.0147	.0116	.0094
	10	1.0496	.2619	.1163	.0654	.0419	.0291	.0214	.0164	.0129	.0105

to within approximately $6'$ of arc, or $R(\tan\frac{1}{10})/2$. These values are shown in Table 5.2 as a percentage of CRT screen height for values of R between 1 and 10.

To put this into perspective, assume device-independent coordinates in the range $-1 \le x \le 1$ and $-1 \le y \le 1$ scaled to a 12-inch vertical height CRT screen, and $d = 5$ units. We will show in the next section that the proper viewing distance from an observer to the stereo window is R. Therefore, an observer would have to sit at least 30 inches from the CRT based on the maximum breakdown limit ($R = 5$), and 48 inches from the CRT based on the maximum refusion limit ($R = 8$).

Shape of the Stereo Window

We will show that rotation with perspective projection produces a non-planar stereo window that distorts an image displayed on a flat screen. This distortion has been shown by numerical and graphical examples by Saunders [SAUN68] and has been interactively demonstrated and described by Butts and McAllister [BUTT88a]. We derive an analytic expression that fully characterizes this distortion.

The horizontal parallax, H, is equal to $x_{lw} - x_{rw}$. Therefore, from Equations (5.1) and (5.3),

$$H = \frac{d[x_0 \, \cos(\phi/2) + (z_0 - R) \, \sin(\phi/2)]}{(z_0 - R) \, \cos(\phi/2) - x_0 \, \sin(\phi/2) + R}$$

$$- \frac{d[x_0 \, \cos(\phi/2) - (z_0 - R) \, \sin(\phi/2)]}{(z_0 - R) \, \cos(\phi/2) + x_0 \, \sin(\phi/2) + R}$$

Table 5.2 Vertical parallax values in which fusion can be maintained and for fusion after breakdown. Values are shown relative to normalized device coordinates for R between 1 and 10. The observer is assumed to be a distance R (scaled to device coordinates) from the CRT screen.

R	Limit of Fusion	Fusion after Breakdown
1	.0058	.0017
2	.0116	.0035
3	.0175	.0052
4	.0233	.0070
5	.0291	.0087
6	.0349	.0105
7	.0407	.0122
8	.0465	.0140
9	.0524	.0157
10	.0582	.0175

To simplify notation let $A = \sin(\phi/2)$, $B = \cos(\phi/2)$, and $C = z - R$. Then, after finding a common denominator, we have

$$H = \frac{d(x_0 B + CA)(CB + x_0 A + R) - d(x_0 B - CA)(CB - x_0 A + R)}{(CB)^2 - (x_0 A + R)^2}$$

$$= \frac{d(2x_0^2 AB + 2C^2 AB + 2CAR)}{(CB + R)^2 - (x_0 A)^2}$$

Since $2AB = 2\sin(\phi/2)\cos(\phi/2) = \sin\phi$, this reduces to

$$\frac{d\left\{\left[x_0^2 + (z_0 - R)^2\right]\sin\phi + 2R(z_0 - R)\sin(\phi/2)\right\}}{[(z_0 - R)\cos(\phi/2) + R]^2 - [x_0 \sin(\phi/2)]^2}$$

At the stereo window, horizontal parallax $H = x_{lw} - x_{rw} = 0$, so we have

$$\frac{d\left\{\left[x_0^2 + (z_0 - R)^2\right]\sin\phi + 2R(z_0 - R)\sin(\phi/2)\right\}}{[(z_0 - R)\cos(\phi/2) + R]^2 - [x_0 \sin(\phi/2)]^2} = 0$$

Setting the numerator to zero, we have

$$d\left\{\left[x_0^2 + (z_0 - R)^2\right]\sin\phi + 2R(z_0 - R)\sin(\phi/2)\right\} = 0$$

which implies

$$\frac{x_0^2 + (z_0 - R)^2 + (z_0 - R)[2R\sin(\phi/2)]}{\sin\phi} = 0$$

Completing the square,

$$x_0^2 + \left[\frac{z_0 - R + R\sin(\phi/2)}{\sin\phi}\right]^2 = \left[\frac{R\sin(\phi/2)}{\sin\phi}\right]^2 \qquad (5.7)$$

which is the equation for a circle in the xz plane with center $(0, 0, R[1 - \sin(\phi/2)/\sin\phi])$ and radius $R\sin(\phi/2)/\sin\phi$. Since the values of ϕ are small, usually $\leq 10°$, $2\sin(\phi/2) = \sin\phi$ to three decimal places. Using this simplification, we have a circle with center at $(0, 0, R/2)$ and with radius $R/2$. Since part of the circle is outside of the view volume, the result is a semicylindrical stereo window. This is shown in Figure 5.5.

On a CRT display the computer screen serves as the window through which we look at a screen. Points with zero horizontal parallax seem to lie in the plane of the (almost flat) CRT screen. A curved stereo window causes distortions in depth relationships in the image. This distortion is demonstrated with the random-dot stereogram shown in Figure 5.6. Random-dot stereograms have been used extensively by Julesz [JULE71] and others to study stereopsis. Monocular cues are removed, and only parallax differences are provided between right- and left-eye views. To create Figure 5.6 we began with a rectangular polygon located in the $z = R/2$ plane. Both the polygon and the background were textured using a random-dot pattern. The random-dot texture of the polygon was then projected onto the background using the rotated perspective projection for generating left- and right-eye views. The stereo image should appear to be a rectangle sitting in front of the CRT screen. Instead, the curvature of the stereo window (in

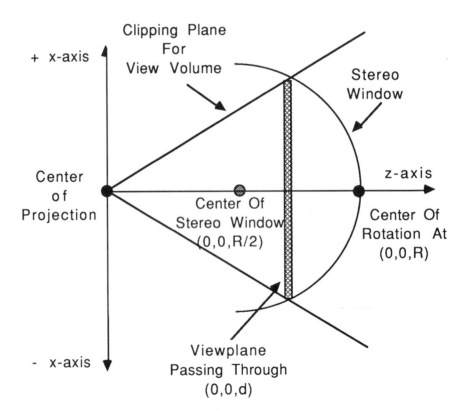

Figure 5.5
Semicylindrical
stereo window.

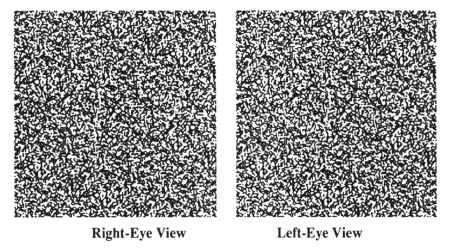

Right-Eye View   **Left-Eye View**

Figure 5.6 Random-dot stereogram showing image distortion produced by rotated perspective projection (arranged for cross viewing).

conjunction with the round-off error resulting from mapping each dot to the nearest pixel location) causes the polygon to break into sections that appear to lie in different depth planes.

Summary for Rotated Perspective

This approach to computing a stereo pair transformation results in vertical parallax between the left- and right-eye perspective views. We have exhibited an analytic expression that characterizes this parallax in terms of the center of rotation, the location of the viewplane, and the angle of rotation. Using this expression, we have shown that the maximum parallax occurs at the corners of the CRT screen and is minimized along the x and y axes of the screen. This expression allows us to choose a center of rotation and viewplane distance that put the maximum vertical parallax within acceptable limits based on the results of Fender and Julesz ([FEND67], [JULE71]).

We have also derived an analytic expression showing that rotations with perspective projection result in a semicylindrical stereo window with center at approximately $(0, 0, R/2)$ and radius $R/2$. Since this semicylindrical window is mapped to the "flat" screen of the CRT, depth relationships can be distorted.

5.4 Parallel Projection

Rotations with Parallel Projection

A variation on the rotational model is to replace the perspective projection with a parallel projection along the z-axis onto the projection plane (orthographic projection). The primary motivation for using parallel rather than perspective projection is that it can be computed faster since no division operation is required. This approach is illustrated in Figure 5.7. In this case the left- and right-eye projections of a point P_0 are

$$x_{rw} = x_0 \cos(\phi/2) - (z_0 - R) \sin(\phi/2) \qquad y_{lw} = y_0 \qquad (5.8)$$

and

$$x_{lw} = x_0 \cos(\phi/2) + (z_0 - R) \sin(\phi/2) \qquad y_{rw} = y_0 \qquad (5.9)$$

Since $y_{lw} = y_{rw}$ there is no vertical parallax. Horizontal parallax becomes $H = (x_{lw} - x_{rw}) = 2(z_0 - R) \sin(\phi/2)$. Setting $H = 0$ locates the stereo window at $z = R$ with positive horizontal parallax when $z_0 > R$ and negative parallax when $z_0 < R$. We have a planar stereo window and have eliminated the vertical parallax. Proper viewing distance is R scaled to device coordinates. However, considerations exist with this technique that did not exist for perspective rotations.

Since parallax varies unboundedly with z, the z-coordinate should be limited to a range such that neither positive nor negative parallax exceeds the interocular distance. In the case of positive parallax, if homologous points in the left- and right-eye views exceed this distance, the observer

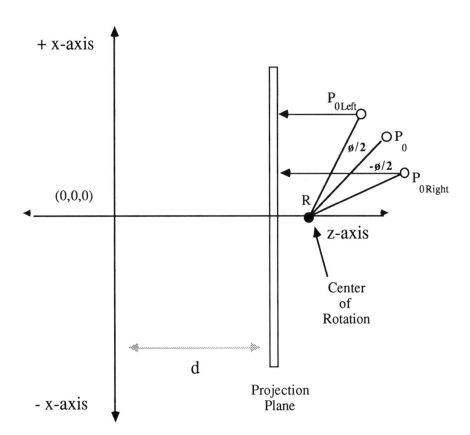

Figure 5.7
Rotation with
parallel
projection.

is forced to look to the left with the left eye and to the right with the right eye; the observer must go wall-eyed. If we assume that the maximum desirable positive horizontal parallax is e device-independent units, then setting $e = 2(z_0 - R) \sin(\phi/2)$ yields an upper bound for acceptable z-values of $z \leq e/[2\sin(\phi/2)] + R$. For the lower bound two conditions must be satisfied. To ensure that maximum negative horizontal parallax does not exceed $-e$, the lower bound for z-values is $z \geq -e/[2\sin(\phi/2)]+R$. To avoid distorting the depth perception outside the CRT screen we also would like this value of z to equal $R/2$ since parallax of $-e$ will appear to put a point halfway between the observer and the stereo window. The center of rotation, $(0, 0, R)$ should be chosen such that $R = e/\sin(\phi/2)$.

Lack of perspective in the image can also result in a reverse perspective effect, in which objects appear to grow in size as they move away from the observer. This method was used for computing stereo pairs of digital terrain models [FORN87]. The effect was noticeable with a single stereo view. It was even more apparent for a series of views for a multiple-perspective three-dimensional image based on Marshall's method [MARS84].

Shear with Parallel Projection

Because the angles used are small, usually $4° \leq \phi \leq 7°$, it has been suggested that the substitutions $\sin(\phi/2) = \phi/2$ and $\cos(\phi/2) = 1$ be used as an approximation to a rotation [LIPS79]. In this case the left- and right-eye projections of a point P_0 are

$$x_{rw} = x_0 - (z_0 - R)(\phi/2) \qquad y_{lw} = y_0 \qquad (5.10)$$

and

$$x_{lw} = x_0 + (z_0 - R)(\phi/2) \qquad y_{rw} = y_0 \qquad (5.11)$$

This approach may be viewed as a z-shear and then a translation of the x-coordinate. Since every x-coordinate is translated by the same amount, this operation could be accomplished by panning the entire right-eye view to the right by $R(\phi/2)$ and the entire left-eye view to the left by $R(\phi/2)$. This simplifies the left- and right-eye projections of a point to simple z-shears of the x-coordinate:

$$x_{rw} = x_0 - z_0(\phi/2) \qquad y_{lw} = y_0 \qquad (5.12)$$

and

$$x_{lw} = x_0 + z_0(\phi/2) \qquad y_{rw} = y_0 \qquad (5.13)$$

The same considerations of unbounded parallax and reverse perspective hold for the shear with parallel projection as for the rotation with parallel projection.

Clipping Plane

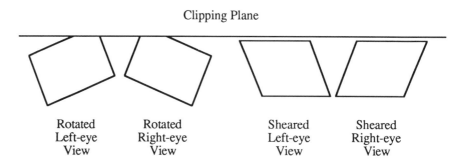

Figure 5.8
Rotations and
shears with
z-clipping plane.

| Rotated Left-eye View | Rotated Right-eye View | Sheared Left-eye View | Sheared Right-eye View |

The shear does, however, offer two advantages. An obvious one is that it requires fewer operations, resulting in less computation time than the rotation. A less obvious advantage is that the shear correctly handles z-clipping planes. It is possible for portions of the image to be clipped for one eye view but not the other when rotations are used, since the z-coordinate of a point P is displaced to $z_r = x \sin(\phi/2) + (z - R) \cos(\phi/2)$ for the right-eye view and to $z_l = -x \sin(\phi/2) + (z - R) \cos(\phi/2)$ for the left-eye view. A shear leaves the z-coordinates of each view unchanged, as shown in Figure 5.8.

Summary for Parallel Projection
When used with parallel projections, both rotations and shears are free of vertical parallax and image distortion. Both methods exhibit unbounded horizontal parallax and can give the observer a perception of reverse perspective, especially in multiview stereoscopic images. Shears are easier to compute than rotations. Shears also have an advantage over rotations in that the left- and right-eye perspective views are clipped identically by a z-clipping plane.

5.5 Two Centers of Projection

Equations for Left- and Right-Eye Views
Another approach to producing the desired binocular disparity between the left- and right-eye views is implemented by choosing two centers of projection for a perspective projection onto a viewplane—one for the right-eye view (RCoP, for *right center of projection*) and one for the left-eye view (LCoP, for *left center of projection*). RCoP and LCoP differ by a distance e in the x-coordinate direction. Figure 5.9 illustrates this model. As before, we assume device-independent screen coordinates with $-1 \leq x \leq 1$ and $-1 \leq y \leq 1$ and the projection plane parallel to the xy plane and positioned a distance d along the z-coordinate axis. LCoP is located at the point $(+e/2, 0, 0)$, and RCoP is located at the point $(-e/2, 0, 0)$.

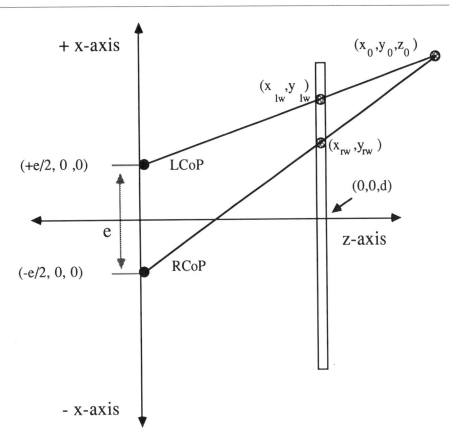

Figure 5.9
Model using two centers of projection.

The line of projection connecting a point (x_0, y_0, z_0) with LCoP is given by the parametric equations

$$x = x_0 + t\,(e/2 - x_0) \qquad y = y_0 + t(0 - y_0) \qquad z = z_0 + t(0 - z_0)$$

At the viewplane $z = d$. Solving for t at this point gives $t = (z_0 - d)/z_0$. Substituting this value of t into the equations for x and y gives the coordinates (x_{lw}, y_{lw}) of the projected point on the stereo window:

$$x_{lw} = \frac{d x_0}{z_0} - \frac{e d}{2 z_0} + \frac{e}{2} \qquad y_{lw} = \frac{d y_0}{z_0} \qquad (5.14)$$

Similarly, the line connecting the point (x_0, y_0, z_0) with RCoP produces a projected point on the stereo window with coordinates

$$x_{rw} = \frac{d x_0}{z_0} - \frac{e d}{2 z_0} - \frac{e}{2} \qquad y_{rw} = \frac{d y_0}{z_0} \qquad (5.15)$$

Parallax

From inspection of the equations for the left- and right-eye views, it is evident that this approach does not exhibit any of the problems encountered with rotations. Since $y_{rw} = y_{lw}$, there is no vertical parallax. Horizontal parallax, H, is represented by the difference

$$H = (x_{lw} - x_{rw}) = \frac{-ed}{z_0} + e \qquad (5.16)$$

When $z_0 > d$, d/z_0 is less than 1, which implies that H is greater than 0 and is bounded by e. Similarly, when $z_0 < d$, $d/z_0 > 1$, which implies $H < 0$, yielding negative parallax. When $d = z_0$, $H = 0$, so that the stereo window is planar and coincides with the viewplane window.

Maximum Parallax Values

We want the projection of e onto the CRT screen to be small enough that the left- and right-eye views can be fused into a single 3D image. Since the device-independent viewplane window (which is two units square) will be scaled to fit the CRT screen (device coordinates) when the image is rendered, all points must project onto the actual viewing screen with a horizontal parallax $H = x_{lw} - x_{rw} = (-ed/z_0) + e$ scaled by the relative width of the screen. For example, left and right perspective views on a 10-inch screen will be physically closer together in absolute measurements than the same views rendered on a 20-inch screen. For this relative parallax to be within the range of the typical observer's ability to fuse the images, we suggest $e = 0.028d$ [HODG92]. To avoid the stretching effect discussed in Section 5.2 the optimal viewing distance from the CRT is equal to d scaled to device coordinates.

Implementation

The two-centers-of-projection approach may also be implemented using a single fixed center of projection and similarity transformations. Using homogeneous coordinates and matrix transformations, we may express the left-eye view in terms of e and d as $x_{lw} = x'/w$ and $y_{lw} = y/w$ where $(x, y, z, 1)\mathbf{L} = (x', y, z, w)$ and

$$\mathbf{L} = \begin{bmatrix} 1 & 0 & 0 & 0 \\ 0 & 1 & 0 & 0 \\ \frac{e}{2d} & 0 & 1 & \frac{1}{d} \\ -\frac{e}{2} & 0 & 0 & 0 \end{bmatrix}$$

The right-eye view may be expressed in terms of e and d as $x_{rw} = x'/w$ and $y_{rw} = y/w$ where $(x, y, z, 1)\mathbf{R} = (x', y, z, w)$ and

$$\mathbf{R} = \begin{bmatrix} 1 & 0 & 0 & 0 \\ 0 & 1 & 0 & 0 \\ \frac{e}{2d} & 0 & 1 & \frac{1}{d} \\ \frac{e}{2} & 0 & 0 & 0 \end{bmatrix}$$

The matrices $\mathbf{L}$ and $\mathbf{R}$ can be decomposed into $\mathbf{L} = \mathbf{T}_l \mathbf{P} \mathbf{T}_l^{-1}$ and $\mathbf{R} = \mathbf{T}_r \mathbf{P} \mathbf{T}_r^{-1}$ where $\mathbf{T}_l(\mathbf{T}_r)$ is a translation, $\mathbf{P}$ is a perspective projection, and $\mathbf{T}_l^{-1}(\mathbf{T}_r^{-1})$ is the inverse translation. Decomposing $\mathbf{L}$ and $\mathbf{R}$, we have

$$\mathbf{T}_l = \begin{bmatrix} 1 & 0 & 0 & 0 \\ 0 & 1 & 0 & 0 \\ 0 & 0 & 1 & 0 \\ -\frac{e}{2} & 0 & 0 & 0 \end{bmatrix} \qquad \mathbf{T}_r = \begin{bmatrix} 1 & 0 & 0 & 0 \\ 0 & 1 & 0 & 0 \\ 0 & 0 & 1 & 0 \\ \frac{e}{2} & 0 & 0 & 1 \end{bmatrix}$$

$$\mathbf{P} = \begin{bmatrix} 1 & 0 & 0 & 0 \\ 0 & 1 & 0 & 0 \\ 0 & 0 & 1 & \frac{1}{d} \\ 0 & 0 & 0 & 1 \end{bmatrix}$$

This approach may also be derived intuitively by considering the geometry. Assume a right-handed coordinate system with center of projection (CoP) at $(0, 0, 0)$ and a viewplane positioned at $z = d$. To compute the right-eye perspective of a point $P_0 = (x_0, y_0, z_0)$, we first translate it along the x-axis a distance $e/2$. Similarly, to compute the left-eye perspective of P_0, we translate it along the x-axis a distance $-e/2$. Thus, for P_0 we have

$$P_{0\text{right}} = \left(x_0 + \frac{e}{2}, y_0, z_0\right) \qquad P_{0\text{left}} = \left(x_0 - \frac{e}{2}, y_0, z_0\right)$$

The line of projection from $P_{0\text{right}}$ to the center of projection is given by the parametric equations

$$x = x_0 + \frac{e}{2} + t\left[0 - \left(x_0 + \frac{e}{2}\right)\right] \qquad y = y_0 + t(0 - y_0) \qquad z = z_0 + t(0 - z_0)$$

At the viewplane $z = d$, which gives $t = (z_0 - d)/z_0$ and

$$x_{rw} = \frac{dx_0}{z_0} - \frac{ed}{2z_0} \qquad y_{rw} = \frac{dy_0}{z_0} \qquad (5.17)$$

Similarly, for $P_{0\text{left}}$ we have

$$x_{lw} = \frac{dx_0}{z_0} + \frac{ed}{2z_0} \qquad y_{lw} = \frac{dy_0}{z_0} \qquad (5.18)$$

Horizontal parallax is represented by the difference

$$(x_{lw} - x_{rw}) = -\frac{ed}{z_0} \qquad (5.19)$$

Note that, if we now display our stereo image with no further transformations, we will obtain crossed parallax since z_0 is always greater than zero. The entire image will be projected in front of the stereo window. Comparing Equations (5.17) and (5.18) with Equations (5.14) and (5.15), we see that to modify the stereo image to match our viewing geometry, we translate the projected point by adding $e/2$ to x_{lw} and subtracting $e/2$ from x_{rw} to get the proper views. If the perspective projection is computed in hardware with a fixed center of projection, the same result may be achieved by panning the left-eye image to the left and the right-eye image to the right.

In graphics systems with hardware translations, perspective projection, and panning, an efficient implementation of the two-centers-of-projection method can be effected by applying $\mathbf{T}_l\mathbf{P}$ (or $\mathbf{T}_r\mathbf{P}$), followed by a pan.

Moving the Stereo Window

It is a standard technique in stereo photography to alter the position of the stereo window by increasing or decreasing the amount of separation between the left- and right-eye views. An increase in separation moves the stereo window forward relative to the image; a decrease in separation moves the stereo window backward relative to the image. This is most easily done on a CRT display with a hardware pan command. It can also be done in software by scaling the constant $\partial(e/2)$ term in the expression

$$x_w = \frac{dx_0}{z_0} - \frac{\partial ed}{2z_0} + \partial\frac{e}{2}$$

where $\partial = +1$ for the left-eye view and -1 for the right-eye view. Assuming a scaling factor m, the horizontal parallax $(x_{lw} - x_{rw})$ becomes $-(ed/z_0) + me$, and the stereo window is shifted to d/m along the z-axis.

Summary for Two Centers of Projection

The computation of stereoscopic views using two centers of projection suffers from none of the artifacts that occurred with rotations. There is no vertical parallax, the stereo window is planar, and the maximum positive horizontal parallax is bounded. In addition, we have shown that this method may be obtained with a single, fixed center of projection using similarity transformations. This allows an efficient implementation using hardware translation, projection, and panning. Also, this method is the starting point for the recent work proposed in [DEER92], which models the screen curvature and the glass thickness to achieve accurate, high-resolution, head-tracked stereo displays on a workstation CRT.

5.6 Summary and Conclusions

We have analyzed the different approaches to computing the left- and right-eye views of a stereo pair for display on a CRT screen. Rotations used with perspective projection result in vertical parallax between the left and right perspective views. We have presented an analytic expression that characterizes this parallax in terms of the center of rotation, the location of the viewplane, and the angle of rotation. We have also derived an analytic expression showing that rotations with perspective projection result in a semicylindrical stereo window with center at approximately $(0, 0, R/2)$ and radius $R/2$. Since this semicylindrical window is mapped to the "flat" screen of the CRT, depth relationships can be distorted. Because of this distortion, this approach is unacceptable for CRT display.

When used with parallel projections, both rotations and shears are free of vertical parallax and image distortion but exhibit unbounded horizontal parallax, and the lack of perspective can give the observer a perception of reverse perspective. Shears are computationally more efficient and have an advantage over rotations in that the left and right perspective views are clipped identically by a z-clipping plane.

Computing stereoscopic views using two centers of projection suffers from none of the artifacts that occurred with rotations. There is no vertical parallax, the stereo window is planar, and the maximum positive horizontal parallax is bounded. In addition, we have shown that this method may be obtained with a single, fixed center of projection using similarity transformations. This allows an efficient implementation using hardware translation, projection, and panning.

6

Liquid-Crystal Shutter Systems for Time-Multiplexed Stereoscopic Displays

Philip J. Bos

6.1 Introduction

Stereoscopic computer displays must provide one view of a computer-generated image to the user's left eye and another to the right eye. This should be accomplished in a way that is compatible with existing monoscopic systems. Further, as Lipscomb [LIPS89] has pointed out, the system should be "unobtrusive" to the user.

To be effective, the device must provide separate images to each eye. It is important that the left eye not perceive the right eye's image and vice versa. Early methods of providing stereoscopic images accomplished this by using two separate displays and some optical means to ensure that only one of the displays could be viewed by each eye. However, these systems deviated considerably from the architecture of existing monoscopic systems and have not been widely accepted.

Field-sequential or time-multiplexed stereoscopic viewing systems can provide stereoscopic images without requiring major changes to existing display systems, and the level of graphics performance is not substantially reduced. Field-sequential stereoscopic systems display time-sequentially the two eye views of an object that the observer would have if the object were actually present. The system also has some means of sorting out the two images to ensure that the correct image gets to the correct eye. One of the original techniques to accomplish this used a mechanical shutter that blocked light from each eye during the time interval when the other eye's image was being displayed. To avoid the perception of this shuttering, the system had to repeat each eye's image at a fast rate (about 60 Hz) [BEAT86].

Although the early mechanical shuttering techniques did work with some existing display systems, they were obtrusive. Lipscomb [LIPS89] calls obtrusiveness "the degree to which a user's attention is sapped from the application to deal with the stereo device" (p. 30). The major causes of obtrusiveness, according to Lipscomb, were the weight of the devices and their restricted field of view.

In an effort to design a field-sequential stereoscopic viewing system that is less obtrusive, electro-optical materials have been used. Although some devices have been designed with PLZT ceramics [ROES76], most systems currently used or discussed employ liquid crystals as the electro-optical material. Lipscomb [LIPS89] has developed a time line of stereoscopic viewing devices at UNC and compared their obtrusiveness. Liquid-crystal (LC) glasses were shown to be less obtrusive than previous field-sequential stereoscopic viewing devices, and a polarizing plate (also liquid-crystal-based) was found to be even less obtrusive. (Subsequent to the compilation of this time line, active glasses using a wireless link have been developed that are even less obtrusive than the "tethered" versions considered by Lipscomb.)

The two LC devices described by Lipscomb are representative of two approaches to the design of a field-sequential stereoscopic viewing system. Either the viewer wears *active glasses*, which house the electro-optical device, or the system is designed with *passive glasses*, where the electro-optical device is attached to the front of the CRT. Both of these approaches have incorporated various designs and liquid-crystal-based electro-optical effects.

In order to understand the operation of these devices, it is necessary to understand a little about liquid crystals and the basic electro-optical effects that have been developed with them. The next section will cover these topics. Subsequent sections will cover LC devices developed for stereoscopic viewing applications, the performance of stereoscopic viewing systems using active glasses, and the design and performance of stereoscopic viewing systems using passive glasses.

6.2 Basic Liquid-Crystal Electro-optical Effects

The Basics of Liquid Crystals

The term *liquid crystals* refers to materials that are in a particular phase of matter. Most materials can be in the solid and liquid condensed phases, but some materials, consisting of molecules that are elongated in shape, can exist in other condensed phases of matter called the liquid-crystalline phases. In the solid, or crystalline, phase, the elongated molecules have two types of order. They are translationally ordered; that is, their centers of mass fall in defined locations relative to neighboring molecules. They are also orientationally ordered; that is, their long axes are aligned relative to neighboring molecules. In the liquid phase, *both* of these types of order are lost. So it is not surprising that at temperatures between the solid and liquid phases, there are phases in which the molecules have orientational order and varying degrees of translational order. These are called the liquid-crystalline phases [PRIE75].

The most widely used liquid-crystalline phase of matter for electro-optical devices is the nematic phase. In this phase the molecules have some orientational order but no translational order. That is, the long axes of the molecules tend to be parallel to each other, but the molecules are not fixed in location relative to their neighbors. Materials that are designed to exist in the nematic phase of matter at room temperature are called nematic liquid crystals. Because of their lack of translational order, nematic liquid crystals can flow like a liquid, but their orientational order gives them some properties of a solid.

Materials in the nematic phase don't exhibit perfect orientational order, though. The long axes of neighboring molecules are undergoing considerable thermal fluctuations, and although the long axes of neighboring molecules are *on the average* aligned along a common axis, they are seldom actually aligned along that axis. The axis that is defined by the local average of the long molecular axes is called the director axis.

In a liquid crystal, there are elastic forces that are minimized when the director is everywhere parallel, but certain factors can affect the director field. One of these factors is surface alignment. By appropriately treating the surface of a material (such as glass) that is in contact with a drop of liquid-crystalline material, the director at the surface can be made parallel to the surface, perpendicular to the surface, or at some intermediate angle.

The other factor to consider is the torque applied to the director by an electric field. The thermal motions of a liquid-crystal molecule about the director average out any permanent electric dipoles but cause the material to have a different polarizability along the director from that perpendicular to the director. If a liquid crystal is placed in an electric field, electric dipoles will be induced in the sample parallel and perpendicular to the director that are proportional to the polarizability along those directions. These dipoles will tend to align along the same axis as the electric field, so if the material has a greater polarizability along the director, it will experience a torque tending to align it along the electric field direction.

In most cases, the polarizability is greater along the director axis. This results in the dielectric constant being greater along the director axis than perpendicular to it, and materials of this type are said to have a positive dielectric anisotropy (or, simply, positive type materials).

Consider what would happen if a nematic liquid crystal were placed between two pieces of glass whose surfaces have been treated so that the directors in contact with a surface will everywhere be parallel, and tipped a couple of degrees out of the plane of the surface. If the two pieces of glass are positioned so that the alignment axes of the two surfaces are parallel to each other, the director field will align in the cell everywhere parallel to that axis. This is because, as mentioned before, certain elastic forces are minimized when the director in one location is parallel to its neighbors.

Now consider the introduction of an electric field to this device through the use of a transparent conductor applied to the inside surfaces of the glass. It turns out that for the magnitude of the torques exerted by the electric field, the effect of the field on the alignment of the directors at the surface can be neglected. The resulting director field in the cell is then defined by the balance between the elastic torque, which is minimized when the director is parallel to its neighbors, and the electric torque, which is minimized when the director is parallel to the electric field (perpendicular to the cell surfaces), under the condition that the director orientation at the surface is fixed by the surface conditioning. The resulting director fields have been calculated [LESL70] and are approximated in Figure 6.1 for low, medium, and high electric field strengths.

This type of liquid-crystal cell can be used to change the polarization state of light passing through it in a manner that is dependent on the director field orientation (which is a function of the applied electric field). This is the basis for the most fundamental type of LC device. This device is referred to by a number of different names, but it is often called an electrically controlled birefringence (ECB) device. To explain the electro-optical effect and how this device could potentially be used as a stereoscopic viewing device, we must first discuss some fundamentals of the optics of liquid-crystal devices.

Some Basic Optical Properties of Liquid Crystals

As mentioned earlier, when an electric field is applied to a liquid crystal, the material is more polarizable if the field is applied along the axis of the director than if it is applied perpendicular to it. The degree of polarizability is dependent on the frequency of the applied electric field, so at optical frequencies the polarizability may not be the same as it is at lower frequencies. However, it turns out that for the materials we are interested

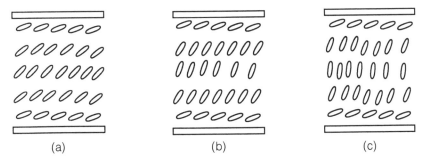

(a)   (b)   (c)

Figure 6.1 • Side view of a liquid-crystal cell where (a) a low-strength electric field is applied vertically (perpendicular to the cell's surfaces), (b) an intermediate-strength field is applied, and (c) a high-strength field is applied.

in, the polarizability at optical frequencies is always greater when the axis of the electric field is along the director axis than when it is perpendicular to the director.

The polarizability of a material at optical frequencies determines the velocity of light passing through it. The polarizability can be expressed in terms of the index of refraction of a material (n), and the velocity of light in a material is given by $v = c/n$, where c is the velocity of light in a vacuum. Because the polarizability, and therefore the index of refraction, of a liquid crystal is dependent on the direction of the electric field relative to the director, a liquid crystal has two indices of refraction and is said to be birefringent [SHUR64].

More specifically, when a ray of light traveling perpendicular to the director is linearly polarized with its electric field axis parallel to the director, it propagates with a velocity $v_e = c/n_e$, where v_e and n_e are called the extraordinary velocity and extraordinary index of refraction, respectively; and if the ray of light is polarized so that its electric field axis is perpendicular to the director, it propagates with a velocity $v_o = c/n_o$, where v_o and n_o are called the ordinary velocity and ordinary index of refraction, respectively. For the case of light traveling at an arbitrary angle to the director and with arbitrary polarization, the electric field vector of the light is broken up into two components. One component is perpendicular to the director and the light propagation direction; this component is called the ordinary component, and the light ray associated with this component is called the ordinary ray. The ordinary ray propagates with velocity v_o. The other electric field component is perpendicular to the ray propagation direction and to the ordinary component. It is called the extraordinary component, and the light ray associated with it is called the extraordinary ray. The extraordinary ray propagates with velocity $v_e = c/n'_e$ where the effective extraordinary index of refraction, n'_e, is given by

$$ n'_e = \left[\frac{(n_e n_o)^2}{(n_e \cos \theta)^2 + (n_o \sin \theta)^2} \right]^{1/2} $$

where θ is the angle between the director and the ray propagation direction. Note that for the case where the light propagation direction is perpendicular to the director, $n'_e = n_e$; and for the case where the light is propagating along the director axis, $n'_e = n_o$.

So when light travels through a liquid crystal, it is broken into two components that travel with different velocities. The velocity of one component is determined by the ordinary index of refraction of the material, and the velocity of the other is determined by the values of the ordinary and extraordinary indices and the angle that the light ray makes with respect to the director axis.

Because the two components travel at different velocities, they have different wavelengths: $\lambda'_e = \lambda_v/n'_e$ and $\lambda_o = \lambda_v/n_o$. As a result, even though these two components enter the liquid-crystalline material in phase, they generally exit out of phase. The phase difference can be quantified in terms of the fractional wavelength that the exiting components are shifted by (say, a half-wave or a quarter-wave) or by giving the corresponding phase angle in radians (a half-wave is $2\pi \times 1/2$ radians; a quarter-wave is $2\pi \times 1/4$ radians). To be more specific, consider a liquid-crystal cell of thickness d. If a linearly polarized light ray hits this cell at an angle perpendicular to its surface, it will be broken up into two components that are initially in phase. Within the cell there will fit d/λ'_e waves of the extraordinary component, so that component will have a total phase angle of $2\pi d/\lambda'_e$. Likewise, the total phase angle for the ordinary component is $2\pi d/\lambda_o$. The difference between these numbers is the phase shift and can be expressed as: $\Delta\phi = 2\pi d\,\Delta n'/\lambda_v$, where $\Delta n'$ is the difference between n'_e and n_o.

To see the significance of this phase shift, consider the specific example shown in Figure 6.2, where the thickness of the cell is chosen to be $d = \lambda_v/2\Delta n'$. In this case $\Delta\phi$ is π radians, and the cell is said to be a half-wave retarder. Linearly polarized light is incident on the cell with its propagation direction perpendicular to the director in the cell and with its instantaneous electric vector at 45° to the director axis. As shown, the initial component is broken up into two components that are initially in phase but on exiting the cell are π radians out of phase. The interesting feature is that when these

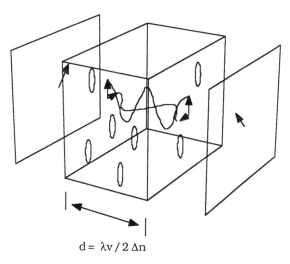

$$d = \lambda v / 2 \, \Delta n$$

Figure 6.2 A half-wave retarder between crossed polarizers. The instantaneous electric field components at the polarizers and surfaces of the retarder are shown as arrows.

components exit the cell, they recombine to form linearly polarized light that is rotated 90° from the input polarization axis. So if a half-wave retarder, such as this cell, is placed between crossed polarizers with its director axis at 45° to the polarizer axis, the assembly will have high light transmission. The expression for the transmission of light through an assembly of two crossed ideal polarizers with a LC cell of arbitrary birefringence between them, aligned as shown in the figure, is $T = \sin^2 \Delta\phi/2$.

An Electrically Controllable Birefringence Device

The basic operation of an ECB liquid-crystal device can now be presented [KANE87]. A cell, as described in the previous section, that consists of two pieces of glass coated with a transparent conductor and an alignment layer, is placed between two crossed polarizers. When a large voltage (around 30 V) is applied to the cell, the director field assumes a configuration like that shown in Figure 6.1c. In this case n'_e is close in value to n_o so that $\Delta n'$, $\Delta\phi$, and therefore T are low. When the voltage is lowered or removed, the director field relaxes to an orientation like that shown in Figure 6.1a or 6.1b. Depending on the thickness of the cell and the birefringence of the liquid-crystal material used (typically $\Delta n = 0.15$), there will be a director configuration where $\Delta n'$ is such that $\Delta\phi$ is π radians, and T will be maximized. This director configuration would be similar to Figure 6.1b if the cell were around 5 μm thick (a typical thickness), would be only a little different from Figure 6.1c for a cell 20 μm thick, and would be more like Figure 6.1a if the cell were around 2 μm thick (a thin cell).

In the next section more detail will be given on these types of devices.

A Twisted Nematic Device

The most common type of LC device is the twisted nematic (TN) device. The cell is built like the ECB cell, but the two glass plates are rotated with respect to each other by 90°. The director configuration as a function of applied voltage has been calculated by Berreman [BERR73]. When no electric field is applied, the director everywhere in the cell is tipped relative to the surfaces of the cell by the angle defined by the surface treatment; and to match the boundary condition of the 90° rotated glass plates, the director uniformly twists about the cell normal by 90° through the cell. When an electric field is applied along the cell normal by transparent conductors on the cell surfaces, the director field approaches a configuration where it is everywhere aligned along the cell normal. Polarizers are applied to both sides of the cell, as for the ECB device, but in this case they are attached so that their transmission axes are aligned or at 90° to the surface projection of the surface-contacting directors.

To understand the operation of the device, consider the director configuration corresponding to a low electric field, and assume that the polarizers are attached with their transmission axes parallel to each other and with the transmission axis of one parallel to the surface projection of the surface-contacting directors nearest it. Light passing through one of the polarizers impinges on the liquid-crystal cell with its polarization direction nearly parallel to the director. If the cell is thick enough (making the twist rate low) that the thickness times the birefringence of the material is large compared with the wavelength of light, the electric vector of the light will follow the twist. In this case, the linearly polarized light entering the cell will exit the cell rotated by 90° and be blocked by the second polarizer. Gooch and Tarry [GOOC75] calculated an exact expression for the light transmission: $T = (\sin^2[(\pi/2)(1 + x^2)^{0.5}], (1 + x^2))$, where $x = 2\Delta n'd/\lambda$.

If a voltage is applied of sufficient strength to align the director in the center of the cell to be parallel to the cell normal, the director field will no longer have a twisted appearance. In this case the directors tend to lie in two planes. In the lower half of the cell the directors lie in a plane defined by the surface-rubbing direction of the lower surface and the cell normal, and in the upper half of the cell they tend to lie in a plane defined by the surface-rubbing direction of the upper plate and the cell normal. In this case, linearly polarized light that impinges the cell as before is transmitted as a purely extraordinary wave in the bottom half of the cell, and as a purely ordinary wave in the top half. As a result, no net phase shifts or rotations of the polarization occur.

To summarize the operation of a twisted nematic device: When no voltage is applied, linearly polarized light is rotated by 90°, so if crossed polarizers are appropriately placed on both sides of the cell, the transmission of the assembly will be high. When a high voltage is applied to the cell, linearly polarized light is unaffected, and the assembly of cell and crossed polarizers will have low light transmission.

6.3 Applications of LC Electro-optical Effects to Active Glasses

Application of the TN Electro-optical Effect

One of the first applications of liquid crystals to stereoscopic viewing devices was described by Roese [ROES77]. He pointed out that twisted nematic devices could be used in active glasses to block light from reaching a user's left or right eye when a right or left view of a scene is displayed on a CRT. Twisted nematic devices can exhibit low light transmission values in the voltage-applied state, so they can effectively block the incorrect image from reaching each eye, but their switching speed is a problem. When an electric field is applied to the device, it can switch to a low transmission value in well under 1 ms; however, when the voltage is removed, the relaxation of the director field to the light-transmitting twisted state typically takes greater than 10 ms.

The problem of speed for TN devices was addressed by Harris and colleagues [HARR86], who applied a technique that was developed by

Raynes and Shanks [RAYN74]. The technique uses liquid crystals called "two-frequency" materials because they exhibit the unusual property that their dielectric anisotropy has a different sign at different frequencies. At low frequencies (say, 100 Hz) this type of nematic liquid crystal has a positive dielectric anisotropy, and the director will experience a torque tending to align it with the the electric field direction. If the frequency is increased above a critical value (say, 5 kHz), the dielectric anisotropy of the material becomes negative, and the director experiences a torque tending to align it perpendicular to the electric field direction. With such devices it is possible to first apply a low-frequency field to cause the shutter to darken quickly, and then, rather than letting the director field simply relax back to its twisted structure, a high frequency field is applied that causes the cell to switch to the light-transmitting state in a shorter time than is usual for TN devices.

The problem of speed was also addressed with the development of a $-3\pi/2$ radian twist cell by Hubbard and Bos [HUBB81]. Dwight Berreman [BERR75] showed that the relaxation of a conventional twisted nematic device is slow not only because the elastic forces driving the relaxation are weak, but more significantly because material flow in the relaxing device produces torques on the director field that oppose the relaxation to the $\pi/2$ radian twisted structure. Berreman further showed that these flow-related torques tend to drive the cell into a $-3\pi/2$ radian twist state. The improved device was designed so that the relaxed state of the device had a three-quarters turn twist rather than the usual quarter turn twist of the director field. In this type of device the relaxation time can be reduced by about a factor of 2 over a comparably built conventional TN device.

Application of the Electrically Controllable Birefringence Effect

ECB devices can also be considered for use as shutters for active glasses. However, this device built with a conventional spacing of around 5 μm has a switching speed problem similar to that of the conventional TN device. Uehara, Mada, and Kobayashi [UEHA75] showed, however, that it is possible to improve the optical response speed by making the device thicker than usual. With the thicker device, Δn needs to change by less to get the desired change (π radians) in $\Delta\phi$. So in a thicker cell the director field needs to move only a slight amount to produce an optical effect, and therefore the optical response time is decreased. Uehara and colleagues showed the fast response of this type of device as it is switched from a voltage V_2 that causes the cell to have a low value of retardation, to a voltage V_1 that causes the cell to have a half-wave retardation. The device they demonstrated used parallel polarizers so that when the cell was in a low retardation state, the light transmission of the cell and polarizers assembly was high; and when the cell was in a half-wave retardation state, the assembly had low light transmission. Similar devices have been reported by Sato and Wada [SATO77], Kwon and colleagues [KWON79], and Fergason [FERG80], who used the term *surface mode* in his description of the device operation. The major drawback of this type of device, as pointed out in [UEHA75], is the limited viewing angle.

The viewing angle of a device is the maximum angle at which the display can be viewed (as measured from an axis perpendicular to the cell's surface) without significant changes being observed in the electro-optical characteristics of the device. Thick devices generally have a small viewing angle because a small change in the angle of the light propagation, even though it results in only a small change in $\Delta n'$, leads to a significant change in $\Delta\phi$.

Another approach to improving the speed of ECB devices while also allowing a large angle of view has been developed by Tektronix [BOS84]. The device, called the π-cell, obtains its speed improvement over usual ECB devices of moderate thickness (around 5 μm) by controlling flow-induced torques on the director field.

Van Doorn [VAND75] first showed that material flow in a relaxing ECB cell, represented by the arrows in Figure 6.3a, imparts a torque on the directors in the central region of the cell that slows the relaxation of the device to the half-wave retardation state. In the π-cell, the surfaces of the cell are treated so that the surface-contacting directors are tilted in opposite directions, as shown in Figure 6.3b. In this type of device, the material flow does not slow the relaxation of the device. For example, a 5 μm thick uniformly aligned ECB device using a liquid crystal with a birefringence of 0.13 relaxes to a state of half-wave retardation in approximately 6 ms, whereas a similarly constructed π-cell will relax in approximately 3 ms.

The viewing angle for a π-cell is good not only because the cell is not excessively thick, but also because of the optically self-compensating nature of the director configuration. In the director configuration of Figure 6.3b, when light strikes the π-cell at an angle to the cell surface such that the effective birefringence in the bottom half of the cell is decreased (because the light propagation direction is more closely aligned with the director axis than for the case of straight-on viewing), the opposite occurs in the upper half, which partially compensates for the change in the bottom half. The off-axis light transmission characteristics of a uniformly aligned ECB device and a π-cell in the half-wave retardation state and between parallel polarizers are shown in Figure 6.4. Although the transmission of the π-cell stays at an acceptably low level, the transmission for the uniform cell is very high.

While the π-cell's switching speed and viewing angle are acceptable for use in active glasses, there can be a problem with obtaining a low light transmission state when crossed polarizers are used. With crossed polarizers,

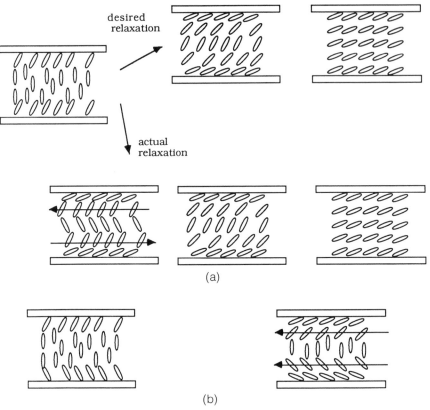

Figure 6.3 The relaxation of the director field in an ECB device after an applied high-strength electric field is removed. The arrows show the direction of material flow in the relaxing cell that can impart a torque on the director near the center of the cell. (a) The torque in a uniformly aligned device is shown to cause the director in the center of the cell to tip in a direction opposite the desired relaxation direction. (b) The flow in a π-cell is shown not to produce relaxation-slowing torques.

the light transmission will be low when $\Delta\phi$ is close to zero, as it will be in the field-applied (or *on*) state, but it is inconvenient to apply the high voltage required to force $\Delta\phi$ to a sufficiently low value. *On* state voltages of convenient levels produce a significant nonzero value of $\Delta\phi$ that causes significant light leakage. This value of $\Delta\phi$ is referred to as the residual retardation of the *on* state, and a method must be employed to correct for it.

An obvious method is to obtain a passive retarder, made from a stretched polymer film, that has a retardation value equal to the residual retardation of the cell in the *on* state. Such a retarder could be placed with its stretch axis, which for on-axis light can be considered optically equivalent to the director

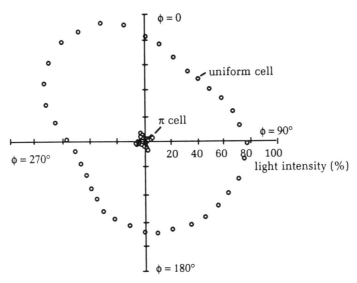

Figure 6.4 The off-axis optical characteristics of a uniform cell and a π-cell in the half-wave state and between parallel polarizers. Light incident at 40° to a perpendicular to the cell's surface was recorded for the azimuthal angles shown. The radial axis is the percent of light transmitted (545 nm) relative to the maximum amount that could be transmitted through the cell and the two polarizers for light directed along a perpendicular to the cell.

axis in a LC cell, aligned so that it is crossed with respect to the projection of surface-contacting directors onto the cell surface. In this case, the net value of $\Delta\phi$ after passing through the *on* cell and the passive retarder will be zero, and the light transmission will be acceptably low. The difficulty with this method is obtaining passive retarders that have the correct value of retardation.

Another method of compensation is due to Jerrard [JERR48]. In this method a passive quarter-wave retarder is used to convert the elliptically polarized light resulting from the residual retardation of the *on* cell to linearly polarized light. The resulting linearly polarized light, however, will have its polarization axis at a small angle to the polarization angle of the first polarizer, so the second polarizer must be tipped to obtain high extinction, as shown in Figure 6.5. Some considerations have been published on the application of this technique to a "surface mode" device by Lipton and Berman [LIPT patent], and to π-cell devices by Bos [B0S93].

Figure 6.6 shows the electro-optical response of this assembly. The ratio of the maximum light intensity measured with no applied electric field to the intensity measured with the field applied is better than 500:1 for this optimized device. It should be pointed out, however, that obtaining this

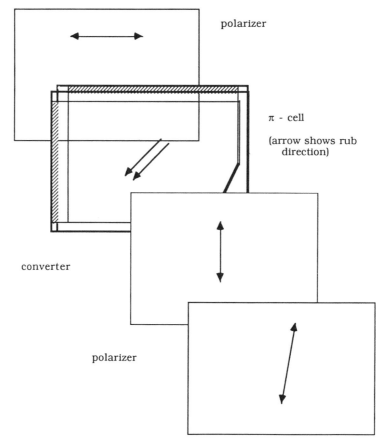

Figure 6.5
An electro-optical device incorporating a π-cell and the Senarmont compensator technique.

very large ratio requires very careful processing of the cell. Ratios of 100:1 to 200:1 are typical for production devices.

Another approach to solving the residual retardation problem that has been applied to active glasses has been reported by Haven [HAVE87]. In this system, which also demonstrates very fast optical switching times, two π-cell-type devices are stacked between crossed polarizers. The cells are

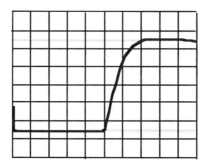

Figure 6.6
The electro-optical response of the device in Figure 6.5. Light transmission is shown on the vertical axis and time (2 ms per division) on the horizontal axis. A high-strength electric field is applied during the first 10 ms, and no field is applied in the second 10 ms.

arranged with the projections of their surface-contacting directors crossed, so that when the two cells are in the same state (*on*, *off*, or relaxing), the net optical retardation of the cells is zero. In operation, both cells are initially *off*, and because the net retardation of the two cells is zero, the device consisting of the two cells between crossed polarizers has very low light transmission. To make the device light-transmitting, voltage is applied to one cell, which quickly attains a low retardation state. The net retardation of the two cells is now just the retardation of one cell in the *off* state, which is chosen to be a half-wave. Thus, the result of applying the voltage to one cell is that the device becomes light-transmitting in much less than 1 ms.

To return the device to a light-blocking condition, voltage is applied to the other cell with voltage still applied to the first one. The result is that, with a switching time again much less than 1 ms, the net retardation of the two cells becomes zero, and the device becomes light-blocking. The system is reset to its original state, of both cells *off*, by simultaneously removing the voltage from both cells. During the relaxation, which takes about 3 ms, the increasing retardation of one cell is always canceled by the increasing retardation of the other, so that the net retardation always remains zero and the device remains light-blocking. Figure 6.7 shows the electro-optical response of the two-cell device driven by a 30 V rms waveform. The ratio of the light intensity of the transmitting state to that of the blocking state is better than 200:1, and the switching time is approximately 50 μs. However, the viewing angle for this device may be smaller than for a single-cell device.

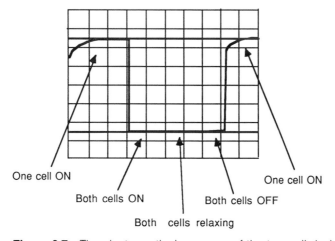

Figure 6.7 The electro-optical response of the two-cell device described by Haven. Light transmission is shown on the vertical axis and time (2 ms per division) on the horizontal axis.

6.4 System Performance of Stereoscopic Viewing Systems Using Active Glasses

The previous sections described liquid-crystal-based devices that seem to be good candidates for the electro-optical element of a stereoscopic viewing system using active glasses. For example, the device shown in Figure 6.5 has been demonstrated to have very low light transmission when voltage is applied to it, and it can switch relatively rapidly to a light-transmitting state. In this section, this device will be used as an example to quantify the performance of a typical active-glasses viewing system.

The performance characteristics of interest are the luminance of the correct image to each eye and the leakage of the incorrect image to each eye (e.g., the right-eye image to the left eye). These characteristics can be quantified by measuring the luminance of a CRT through a LC device that is, like each lens of a pair of active glasses, cycling between low and high transmission states. To obtain data relative to the correct-image luminance, the CRT is driven to full brightness during the time interval that the shutter is open, and it is not driven during the interval when the shutter is closed. For the incorrect-image luminance data, the CRT is driven only during the time interval that the shutter is closed. Because the CRT, during the time interval it is being driven, will scan from the top to the bottom of the screen, the measured characteristics will be a function of the screen position.

To make the raw data values easier to interpret, it is convenient to renormalize them. It is helpful to divide the correct-image data by their peak values to form relative luminance data, so that the brightness setting of the CRT is unimportant. The perception of the incorrect image in a stereoscopic display system is related to the correct-image luminance, so the incorrect-image data will be divided by the correct-image data to form system leakage ratio data.

Figure 6.8 shows relative luminance data and system leakage ratio data for an example of an active-glasses stereoscopic viewing system. This data were acquired for a system where the video scan for each field required 8 ms and where successive fields were separated by a 1 ms blanking interval. This means that the time from when the top of each eye's image is scanned to when the bottom is scanned is 8 ms, and the time from when the bottom of one eye's image is scanned to when the top of the other eye's image is scanned is 1 ms. Because the physical size of the display screen is not relevant to the considered performance characteristics, it is convenient to express the location of a measurement point as the time from the beginning of video, rather than as the distance from the top of the screen.

In considering the performance of the system, the nonuniformity of the relative luminance and the system leakage ratio values from the top to the bottom of the screen is noticeable. To better understand the causes of these variations, it is helpful to consider the data taken from the cell

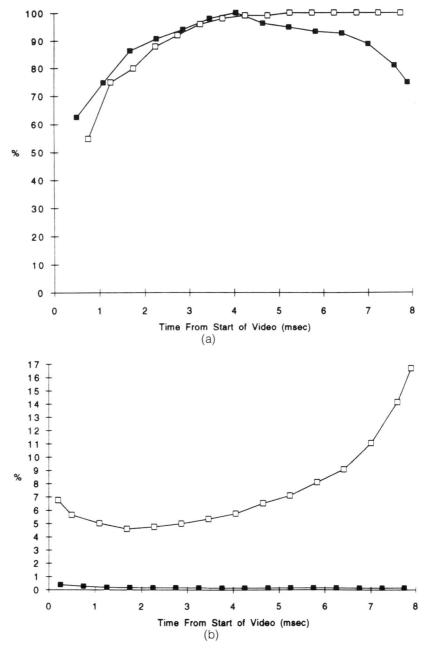

Figure 6.8 (a) The relative luminance of an active-glasses system as a function of screen position (filled boxes) and the relative transmission of the cell alone (open boxes). (b) The system leakage ratio of the same system (open boxes) and the cell leakage ratio (filled boxes). The screen position (from top to bottom) is expressed in terms of the time that position is scanned relative to when the top line is scanned. For these data a color CRT with a standard P22 phospher set was driven to display full white fields (left or right). The LC shutter was driven by a 15 V rms waveform that was applied and removed at the midpoint of successive blanking intervals. The blanking intervals were 1 ms in duration, and the video lasted for 8 ms.

alone. Figure 6.8a also shows the relative transmission of the shutter used, which is defined as the transmission of the shutter during the interval when it's in its high-transmission state divided by its peak value. Figure 6.8b plots the shutter transmission during the time interval when it's in its low-transmission state divided by the shutter's transmission at the same relative time when it is in its high-transmission state. This is known as the cell leakage ratio (CLR).

If the light from the CRT would decay in a time that is insignificantly small compared with the scanning time of the image, the graphs in the two parts of Figure 6.8 would be identical. The extent of their similarity demonstrates the extent to which the electro-optical characteristics of the LC device control the system performance characteristics. It can be seen from the points near the beginning of video (near the top of the screen) that the time required for the LC device to switch to its high-transmission state is the cause of the nonuniformity in the image luminance.

However, in many respects the curves in Figure 6.8a and b are dissimilar. This results from the persistence of the light emission from the phosphor for a significant amount of time after a point has been scanned. The phosphor decay characteristic of a white image of a standard color CRT using the P22 phospher set can be quantified as the amount of light that is emitted in successive milliseconds after scanning (normalized so that the sum of the luminances is 100). From measurement of the area under the luminance versus time curve for successive milliseconds, taken for a spot on a CRT being refreshed at a 60 Hz rate, the results are 76, 8, 4, 3, 2, 2, 1, 1, 0, 0, 0, 0, 1, 1, 1, 0, 0. It can be seen that about 16 percent of the total light flux from a point on the screen is emitted in the time interval beginning 2 ms after the point is scanned. This explains the fall-off in the relative luminance curve at the bottom of the screen; the shutter does not stay open long enough to capture as high a percentage of the light emitted from points near the bottom of the screen compared with points near the center or top. It also explains the lowering of the system leakage ratio at the bottom of the screen. Even though the shutter has a very low transmission when points of the incorrect image near the bottom of the screen are scanned, the shutter is switched to a high transmission state 1 or 2 ms later, and a significant percentage of the light emitted from those points is collected by the incorrect eye.

In summary, the nonuniformity in the image luminance in the top half of the screen is controlled by the shutter's switching speed, and the degree to which the incorrect image is perceived is controlled by the decay characteristics of the CRT's phosphor.

Finally, as pointed out in the introduction, it is important to consider issues related to obtrusiveness. With active glasses four issues come to mind. The first is the weight and obvious obtrusiveness of the glasses themselves, especially for users who wear corrective lenses. The second is the method by which the glasses are "connected" to the CRT's display controller. The third is the low light transmission of the device in association with lighting from the room around the CRT system. The fourth is the strobing effect from viewing CRTs that are not synchronized with the CRT being used for the stereoscopic viewing system.

Although each of these forms of obtrusiveness can be bothersome to a user, they all have been minimized by currently available systems. The first two issues can be considered together. Glasses connected by a wire to the CRT monitor can be made quite light, but the wire is a nuisance; and though the wire can be eliminated with a wireless link, the receiver and batteries mounted in the glasses add weight and volume to them. At the current time, the best compromise seems to be to eliminate the wire and to minimize the receiver weight by using low-power surface-mount electrical components.

The latter two issues are related to the shuttering of the lenses. Because each lens is open only 50 percent of the time, the apparent luminance of objects other than the CRT viewed by the user of active glasses will be 50 percent of the luminance perceived by a wearer of passive polarized glasses, and about 15 percent of the luminance perceived by a person not wearing any glasses. This shuttering also causes a strobing effect when the user views CRTs other than the one used for the stereoscopic display. More specifically, a black bar is seen drifting through the image presented by nonstereo CRTs. Both of these issues can be partially solved by providing a button that the user can push to stop the shuttering of the active glasses when he or she wants to look at objects and other CRTs. More recently, a direction-sensitive IR receiver has been used in place of the button. When the viewer looks away from the stereoscopic display, the shuttering of the device is automatically stopped. Designs for the electronic circuit for active glasses have been described by Lipton and Ackerman [LIPT90] and Prince [PRIN Patent]. An evaluation of active glasses has been done by Beaton [BEAT91].

6.5 Design and Performance of Stereoscopic Viewing Systems Using Passive Glasses

Although a stereoscopic viewing system that uses active glasses requires a minimum of hardware, it has some limitations. From a performance viewpoint, the nonuniformity in the luminance of the correct image and the contribution to incorrect-image intensity due to phosphor persistence are difficult problems to address. And despite careful design to minimize the obtrusiveness of an active-glasses system, the weight of the electro-optical elements, the electronics, and the batteries will probably never be insignificant. To address these limitations, stereoscopic viewing systems using "passive" glasses have been proposed.

Stereoscopic viewing systems using passive glasses have been proposed by Mash and colleagues [MASH86] and Byatt [BYAT81], as well as many others. Their systems use a common twisted nematic device on the front of the CRT, as shown in Figure 6.9. The TN device controls the polarization of light leaving the CRT so that, for example, light leaving the cell during the time interval that the left-eye image is being scanned is horizontally linearly polarized, and light leaving the cell during the interval when the right-eye image is being scanned is vertically polarized. The viewer then wears polarized glasses that transmit only horizontally polarized light to the left eye and only vertically polarized light to the right eye.

A problem with this system is the switching speed of TN devices. For systems where each eye's image is refreshed at a 60 Hz rate, the time for one video scan will be about 8 ms, so a device is needed that can switch in a small fraction of this time. Unfortunately, "fast" TN devices require about 10 ms to switch to the relaxed state.

Many of the other electro-optical effects considered for active glasses have also been considered for a system using passive glasses (for example, the "surface-mode" device [LIPT88]), but because of its acceptable switching speed and angle of view, the π-cell is currently the most commonly used and will be used here as an example.

Basic Design Considerations of a Passive-Glasses Stereoscopic Viewing System

The π-cell will be used to examine the basic design issues of a stereoscopic system with passive glasses [BOS89]. Unfortunately, there is a problem with making a direct substitution of the TN device of Figure 6.9 with

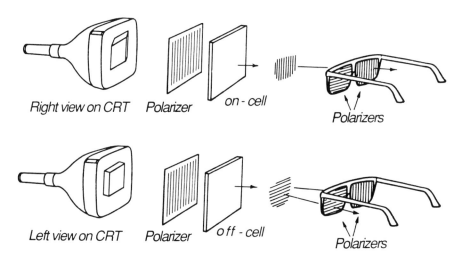

Figure 6.9 A passive-glasses steroscopic viewing system using a twisted nematic liquid-crystal device. The right and left views of an object are alternately displayed on the CRT.

an ECB device such as a π-cell. When an ECB device is in its half-wave retardation state, it is capable of effectively rotating linearly polarized light by 90°, but for only one wavelength of light. The result is that although the light transmission is very low for green light (the color for which the device was designed to be a half-wave retarder), the extinction for blue and red light is not so good [BOS89]. A stereoscopic viewing system that uses passive glasses must address this problem.

The approach to solving this problem is to consider the optical pathway of each eye and to ensure that the net retardation along each of those pathways can be made zero. Figure 6.10 shows an example of a system where this has been accomplished. Note that during the time that the right-eye image is displayed on the CRT, the combined effective retardation of the LC cell (in a zero retardation state) and the crossed passive retarders in the left eye's optical pathway is zero. So, considering the optical components along the left eye's optical pathway, it can be seen that there is effectively nothing between the crossed polarizers, and light of all wavelengths is blocked. Also note in the figure that during the interval that the left-eye image is displayed on the CRT, the combined effective retardation of the LC cell (in a half-wave retardation state) and the two passive quarter-wave retarders that are each crossed with respect to the cell is also zero. So, considering the optical components along the right eye's optical pathway, it can be seen that there is effectively nothing between the crossed polarizers, and light of all wavelengths is blocked.

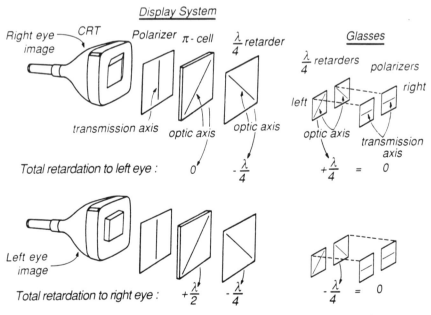

Figure 6.10 A one-cell stereoscopic viewing system using passive glasses.

To demonstrate the improvement of this type of design, the CRT of Figure 6.10 can be replaced by a continuous light source, and the transmissivity of the right eye's optical pathway can be measured as the LC device is cycled between zero and half-wave retardation states. Data acquired in this manner are shown in Figure 6.11, where it can be seen that during the time interval V_2 when the LC device is in its half-wave retardation state, all colors are well extinguished.

An example of a system based on the same principle that uses two LC π-cell devices is shown in Figure 6.12. In this case the two π-cells are arranged with their optic axes crossed, and they are alternately switched between states of zero and quarter-wave retardation. The result is that, in alternate time intervals, the net retardation along each eye's optical pathway is zero, and light of all wavelengths is extinguished by the crossed polarizers.

If the values of the passive and active optical retarders along an optical pathway do not add to zero, the resulting light transmission will be given by the previously considered formula: $T = \sin^2(\Delta\phi/2)$, where $\Delta\phi$ is the net phase shift (in radians) introduced by the birefringent materials along the pathway.

In both of these systems (one- and two-cell), note that when light is to be transmitted along an optical pathway, the net retardation along that pathway is nominally a half-wave. Because of this, the transmission of different colors will not be exactly the same. The difference can be obtained from

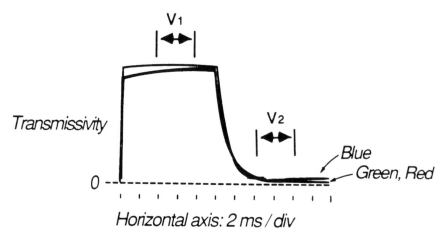

Figure 6.11 The transmissivity of the right eye's optical pathway, shown on the vertical axis, for the system of Figure 6.10. Time is shown on the horizontal axis (2 ms per division). A high-strength electric field is applied during the first 10 ms. During the time interval V_1 the cell is in a low retardation state, and during the interval V_2 it is in a half-wave retardation state.

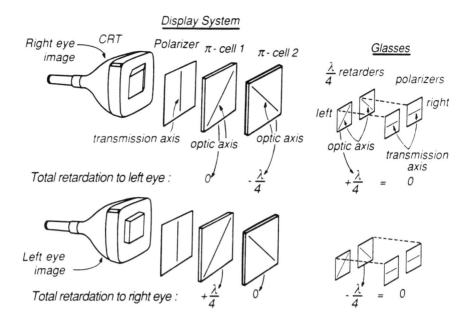

Figure 6.12
A two-cell stereoscopic viewing system using passive glasses.

the preceding formula if the total phase shift is computed for the case of the LC cell having a half-wave retardation (π radian phase shift) for the left eye's optical pathway, and having zero retardation for the right eye's optical pathway. For the purposes of this calculation, it is reasonable to consider the Δn of each retarder as having no dependence on wavelength (this will be discussed more later), and the transmission of blue and red light is found to be about 90 percent of the transmission of green light (assuming the system was designed to maximize the transmission for green light). These differences in transmission are generally not noticeable, but they could be corrected for by readjusting the color balance of the CRT.

Another basic design consideration that can be addressed with a passive-glasses system has to do with phosphor decay. In the case of active glasses not much can be done about this problem, but in a passive-glasses system it is possible to electrically partition the LC cell at the CRT into sections that can be switched independently to allow more time for phosphor decay. For purposes of illustration, consider the case of a shutter that switches instantaneously and has perfect extinction. Figure 6.13 shows for an unsectioned shutter (or active glasses) that at a point near the bottom of the screen, the system leakage ratio is limited to 26.5 percent for a phosphor with the decay characteristics shown. The reason for this high leakage, as discussed in the section on active glasses, is that the shutter switches to allow light to pass to the left eye while the right eye's image is still glowing (and vice versa). As indicated in the figure, 21 percent of the light from the right eye's image is actually transmitted to the left eye.

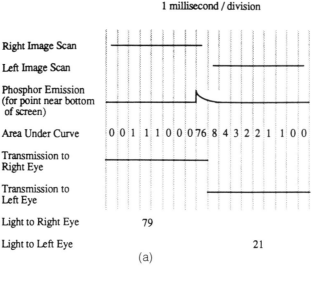

(a)

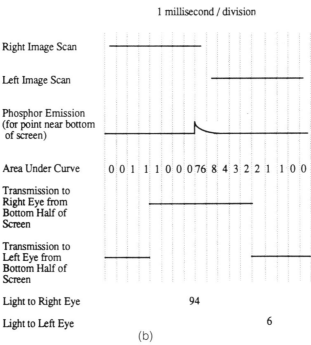

(b)

Figure 6.13 Timing diagram for (a) a single-section and (b) a two-section stereoscopic viewing system showing the effect of phosphor persistence on the incorrect-image intensity. The diagrams show the time when the left and right images are scanned, the light emission from a point of the right image near the bottom of the screen, the times when the optical pathways are open to the left and right eyes, and the amount of light emitted by the considered point of the right image that is collected by the right and left eyes.

Figure 6.13b shows a timing diagram for a shutter electrically divided into top and bottom sections. In this case, the bottom section of the shutter can continue to send light from the lower half of the CRT to the right eye while the top half of the left eye's image is being scanned on the CRT. With the lower section able to send light to the right eye for an additional 4 ms, the light from the right eye's image that is sent to the left eye is decreased to 5.8 percent.

Dividing the cell into more sections further decreases the leakage due to phosphor persistence. If a five-section cell is used, the amount of light from the right-eye image sent to the left eye is decreased to 3.6 percent if the timing of Figure 6.13 is used.

The fact that real shutters have a distinctly noninstantaneous switch to the half-wave state modifies the amount of light leakage due to phosphor decay. The available time from when the image behind a section is finished being scanned at the bottom until it is begun to be scanned at its top is partially used by the time required for the shutter to switch. For a shutter that switches in about 3 ms to a state of perfect extinction, the leakage ratio for the left eye is similar to the numbers just given, but the leakage ratio for the right eye is increased to about 10 percent for a two-section cell and to about 7 percent for a five-section cell.

In summary, the basic design considerations for a stereoscopic viewing system using passive glasses are to provide a net zero retardation along each optical pathway and to section the device to minimize the effect of phosphor persistence.

Performance Characteristics of a Passive-Glasses Stereoscopic Viewing System

To consider system performance issues, the one-cell system discussed in the previous section will be used as an example.

Beyond "basic" system design considerations, the residual retardation of an *on* cell needs to be taken into account. Because of this residual retardation, the LC cell will not have exactly zero retardation when the left eye's image is being presented on the CRT. To ensure that the net retardation along the left eye's optical pathway is zero when the LC cell is *on*, the values of the two passive retarders in the left optical pathway need to be modified by increasing the value of the one at the CRT and/or decreasing the value of the one in the glasses, so that the difference in their retardation is close to the residual retardation of the LC cell.

Figure 6.14 shows the system leakage ratio for the left- and right-eyes views of a passive-glasses viewing system. For this example, the cell has been electrically divided into two sections, and the same phosphor, cell timing, and measurement method as for Figure 6.8 are used. The contribution to the system leakage from the leakage of the shutter is also shown. Although the leakage along the left eye's optical pathway is seen to be quite

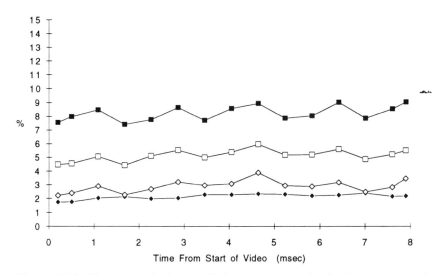

Figure 6.14 The system leakage ratio for a stereoscopic viewing system such as the one shown in Figure 6.10, using a two-section π-cell (boxes). The cell leakage ratio is also shown (diamonds). The data for the right eye are shown as open boxes or diamonds, and the data for the left eye are shown as filled boxes or diamonds. The screen position (from top to bottom) is expressed in terms of the time when a certain position is scanned relative to when the top line is scanned.

low, the leakage along the right eye's pathway shows a variation with time (and so across the screen from top to bottom). This variation is due to the fact that the optical retardation of a π-cell in its *off* state is somewhat time-dependent. This time dependence results from the continuing relaxation of the director field through the configuration that results in a half-wave retardation for green light.

The remaining differences between the cell leakage ratio curves and the system leakage ratio curves are due to the contribution of phosphor persistence. To reduce the system leakage, a cell divided into more sections can be considered. The use of more sections reduces the contribution to the light leakage due to phosphor persistence, as discussed in the previous section, and reduces the variation in the right-eye view leakage due to the time variation of the cell's retardation in the half-wave state. Figure 6.15 shows the system leakage ratio for both views of a system like that of Figure 6.14, but where the cell is divided into five sections.

Another performance issue is the uniformity of the luminance of the correct image across the screen from top to bottom. In the passive-glasses system, the luminance uniformity is improved over that for active glasses. With a two-section shutter the luminance at no point on the screen is less than 80 percent of that of the brightest point on the screen.

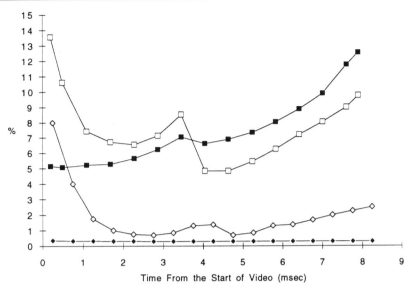

Figure 6.15 The system leakage ratio of a stereoscopic viewing system such as that shown in Figure 6.10, using a five-section π-cell (boxes). Also shown is the system leakage ratio of a stereoscopic viewing system using a five-section π-cell and a color CRT with a rare-earth green phosphor (diamonds). The data for the right eye are shown as open boxes or diamonds, and the data for the left eye are shown as filled boxes or diamonds.

6.6 Future Improvements

As pointed out previously, the persistence of the phosphor is a major contributor to the leakage of the incorrect image to each eye in field-sequential stereoscopic displays. In the future, systems using rare-earth phosphors for the green color component may significantly lower the system leakage ratio.

The problem with current phosphors is related not so much to what is frequently characterized as the decay time (the decay time to 10 percent peak intensity for a white image displayed using the P22 phosphor set is about 350 ms) as to the low-level image persistence, which lasts a considerably longer time. This characteristic is typical of the zinc sulfide phosphors used for the blue and green colors in a standard P22 color phosphor set.

For rare-earth phosphors, the time required for the light emission to decay to 10 percent of its peak value is actually greater than for a zinc sulfide phosphor, but the low-level persistence is less significant. If P43, a rare-earth green phosphor, is substituted for the usual P22 green phosphor in a color CRT, a significant difference in the phosphor decay characteristic of a white image is observed. The luminance measured in successive milliseconds after the phosphor is excited (normalized so that the sum is 100) is 83, 12, 2, 1, 0, 1, 0, 0, 0, 0, 0, 0, 0, 0, 0, 1, 0. In this case, during the first 4 ms 98 percent of the total light is emitted (as compared with only 91 percent for a white image displayed using the standard P22 phosphor set). System leakage data from a color tube using a rare-earth green phosphor are also displayed in Figure 6.15.

Two possible areas of improvement in future viewing systems using active glasses concern the light transmission of the system and, to a lesser extent, the switching speed of the electro-optical element. These improvements will be significant for systems using active glasses because the transmission of room light for such a system is 50 percent of that for a passive system (due to shuttering), and the switching speed is partially responsible for the observed luminance nonuniformity across the displayed image.

As for the first goal, with the systems described so far the maximum light transmission is limited by the transmission achievable with a pair of parallel polarizers (about one-third of the open-gate transmission). Therefore, future active-glasses viewing systems may incorporate electro-optical effects that do not require polarizers.

One such system has been demonstrated by Milgram and Van der Horst [MILL86]. This system uses a liquid-crystal-based electro-optical effect that is capable of switching between a nearly transparent state and a light-scattering state that has optical characteristics similar to a piece of frosted glass. This is called the C-N (cholesteric-nematic phase transition) effect. The nematic liquid crystal used in this type of cell is doped with a chiral molecule that causes the director field to adopt a highly twisted structure when no electric field is applied. As light passes through the cell in this highly twisted state, it does not follow the pitch of the twist and so encounters a changing index of refraction that is responsible for the scattering. When voltage is applied to the cell, the director becomes uniformly aligned along the electric field direction and no longer scatters light.

More recently, new light-scattering liquid-crystal devices have been developed. The PDLC device [ZUME86] at Kent State and the NCAP device [FERG85] developed at Talic utilize microscopic droplets of nematic liquid-crystalline material dispersed or encapsulated in a polymer matrix. When a voltage is applied to one of these devices, the directors inside the microdroplets align along the electric field direction (which is parallel to the propagation direction of light passing through the device). In this case light passing through the droplets does so with the ordinary velocity, and if the polymer matrix has an index of refraction similar to the ordinary index of the liquid crystal used, the medium will be optically homogeneous, and light will pass through it unscattered. If the voltage is removed from the device, the directors in the microdroplets will align in directions not parallel to the light propagation direction. In this case light will propagate with a different velocity in the microdroplets than in the polymer matrix and will be scattered.

Although these light-scattering devices do not have the absorption losses of polarizers in the light-transmitting state, there could be problems with using the light-scattering effect to block the incorrect image. The problem could result from the fact that each of the observer's eyes will see a blank white field instead of a black field for the 50 percent of the time that the incorrect image is being displayed. This could result in an effective lowering of the image contrast ratio. (This problem may be alleviated, however, by using the system in a darkened room or by adding light-absorbing dye molecules to the liquid crystal or polymer matrix.)

Another potential future improvement for active-glasses systems is to decrease the switching time of the electro-optical element for the purpose of increasing the luminance uniformity of the transmitting state as a function of time. Haven, as mentioned earlier, has proposed one method for achieving this. Another would be to use a ferroelectric liquid crystal.

Unlike the liquid-based electro-optical effects discussed previously, ferroelectric devices do not use liquid crystals in the nematic phase. These devices use liquid crystals that are in the more ordered smectic C phase of matter, where the molecules tend to lie in layers with their director axis tipped at an angle to a perpendicular to the planes of the layers. If a chiral additive is mixed with these smectic C materials, the symmetry of the thermal fluctuations of the molecules is changed so that dipole moments associated with the molecules are not completely averaged out in an appropriately aligned bulk sample. The resulting dipole moment of the phase is in the plane of the layers and along an axis perpendicular to the director. From a device viewpoint, this dipole allows the rotation angle of the director about a perpendicular to the planes to be easily controlled by an electric field applied along the plane of the layers. Because this chiral smectic C phase has a dipole moment, liquid crystals in this phase of matter are called ferroelectric liquid crystals.

Using ferroelectric liquid crystals, Clark and Lagerwall [CLAR83] showed how to design an electro-optical device called a surface-stabilized ferroelectric liquid-crystal (or SSFLC) device. The SSFLC device consists of some ferroelectric liquid-crystal material aligned so that the smectic layers are perpendicular to the two plates of glass that contain it. An electric field applied by means of a transparent conductor on the glass plates can cause the dipole moment associated with the material to align "up" or "down" in the cell (depending on the polarity of the applied field). Rotating the dipole causes the director to rotate to two corresponding orientations, where in each case the director is roughly in the plane of the cell but at a different angle about an axis perpendicular to the planes of the layers. The angle between the director in these two states can be made to be about $45°$.

To get the desired electro-optical effect, the cell is made of a thickness such that it is a half-wave retarder and is placed between crossed polarizers aligned so that their axes are perpendicular or parallel to the director (the optic axis) in one of the two states of the device (for example, the "down" state). In the "down" state, light passed by the lower polarizer will pass through the cell purely as an extraordinary wave. The light will therefore not have its polarization state affected as it passes through the cell, and it will be absorbed by the upper polarizer. In the "up" state, the director is rotated at 45° to the polarization axis of the incoming light. As shown in Figure 6.2, for this case light effectively has its polarization axis rotated by 90° as it passes through the cell, and light exiting the cell will be passed by the top polarizer.

Devices built in this way have demonstrated cell leakage ratios of less than 1 percent and can switch in less than 100 μs from one state to the other. These characteristics make them good candidates for an electro-optical element that will improve the luminance uniformity in active glasses. Active glasses that use a SSFLC electro-optical element have been demonstrated by several researchers, including Hartman and Hikspoors [HART87]. The current drawbacks of the SSFLC technology are that the cells are very thin (about 2 μm) and the orientation of their smectic planes can be destroyed by mechanical shock or pressure on the cell plates.

7

Implementation Issues in Interactive Stereo Systems

Louis Harrison

David F. McAllister

7.1 Introduction

As stereoscopic graphics has become cheaper and computing devices faster, commercial interactive stereo systems have begun to appear. We will share our experience in the design of interactive stereo systems, with the discussion geared toward system designers and implementors who wish to produce stereo applications. We will also discuss some of the input devices that are available. Some of these devices are simple and have to be adapted or combined to handle 3D input, and others are designed specifically for that purpose.

We examine two stereo applications, one written before hardware stereo support was available and another, written more recently, that takes advantage of hardware support. Both of these applications allow the user to interactively modify a scene in stereo by drawing new objects, selecting and moving objects, and deleting or erasing objects. This type of application differs from traditional medical imaging, which generally allows only interactive rotation of a scene. These descriptions contain information on cursor selection and other interface issues. Perceptual problems and some commonsense approaches to their treatment are also presented. We explore issues in menu design for stereo. We discuss ways to reduce the cost of rendering by using information from one eye view to help reduce the computation required in rendering the other eye image. We also give an example showing that some algorithms that work well in monoscopic environments may produce anomalies when applied to stereo.

7.2 Stereo Input Devices

Keyboards, joysticks, mice, and trackballs are limited in their ability to input depth information interactively. Some new devices are available to ease the specification of depth and orientation in an interactive stereo environment. We report on some experiences with several of these devices.

The devices will be compared in terms of their input *degrees of freedom*. For the user to move objects interactively and arbitrarily in a 3D view volume, 3D input devices must be able to perform two basic operations: translation and rotation. Each of these operations must be specifiable with respect to any of the three axes. If cursor movement can be translated to one axis at a time, then we say the input device has only one degree of freedom. If the cursor can move with respect to two axes simultaneously, the input device has two degrees of freedom. If a cursor can be translated in any direction through pressure on or movement of the device, then the device has three degrees of freedom. Also, if rotations about any combination of the x, y, and z axes can be specified at the same time as the three types of motion, then we say the device has six degrees of freedom.

Some input devices require the user to sweep a device through space while a sensor determines the position and orientation of the device. The space is called the *draw volume,* and we will compare the accuracy and draw volume size of devices as appropriate. For an input device that is stationary, there is no draw volume.

Keyboard

The keyboard was commonly used as a 3D input device before more sophisticated devices were developed. Keyboards are limited to one degree of freedom for input since only the action of a single keystroke can be recognized at any one time.

Joystick

A joystick control usually moves in eight different directions. In addition to movement in both directions along each of the x and y axes, the joystick can move in both directions along the oblique angles corresponding to the lines $y = x$ and $y = -x$. Although the joystick may appear to be a two-degree-of-freedom input device, it really only has one degree of freedom, since it can move in only one of those eight directions at a time. A key on the keyboard or buttons on the joystick can be used to toggle from rotations to translations, and another key or button can be used to toggle from the xy plane to the xz plane. There are some joysticks that are true two-degree-of-freedom devices, and they are used the same way as mice and trackballs.

Single-Button Mouse

Because the movement of a mouse is restricted to a plane, the mouse has at most two degrees of freedom at any time. Translations are straightforward. Using a key on the keyboard or a button on the mouse to toggle between the xy and xz planes permits the user to move a cursor to any position in a view volume. Curvilinear motion is not possible since drawing must be toggled between the two planes. Using another key on a keyboard to toggle between translations and rotations allows rotations to be accomplished about two axes simultaneously. Usually these are the x-axis and the y-axis. With practice, a user can accomplish rotation about any axis, but this usually requires multiple steps and may not be intuitive for a novice user. For example, a simple rotation about the z-axis requires three steps involving changes in coordinate systems. First a y-axis rotation is performed to rotate the z-axis until it is aligned with the x-axis. Second, the desired rotation is performed as an x-axis rotation. Third, the z-axis is rotated back to the original position using a y-axis rotation.

Multibutton Mouse

The multibutton mouse has additional buttons that can be used to toggle translation between the xz and xy planes. Only two degrees of freedom are available at any time, just as for a single-button mouse, but switching translation axes and the direction of rotation can be accomplished without the user having to direct his or her attention to another input device.

Trackball

The sensing device of a trackball is a sphere that rotates freely in a stationary platform. It is similar in function to a mouse, and it has at most two degrees of freedom. Rotation about any axis in a plane can be specified by spinning the ball about the axis of interest, or translation within a plane can be specified in a way similar to that used with a mouse. Many trackballs also have function keys that can be programmed to toggle between rotation and translation or to toggle between the xy plane and the xz plane. Trackballs have been used as substitutes for mice in portable microcomputers and in applications where there is insufficient desk space for a mouse.

Force and Torque Converters

A force and torque converter is an input device with six degrees of freedom. It consists of a ball mounted on a post with internal sensors that measure torque, or pressure and pressure direction, so that a selected object such as a cursor can be rotated about any axis and translated in any direction simultaneously. A button may be mounted in the ball at the top. Its function can be specified by the programmer, but it is most often used for selection. Function keys may also be mounted on the device so that the user can select actions to be performed without having to direct the eyes away from the screen to locate another input device. The action of the function keys can also be controlled by the programmer, as can the sensitivity of the device. With practice, the user can drive a cursor over a curvilinear path with a high degree of accuracy. Although it is possible to apply both torque and translation vector pressure, most users find it difficult to apply torque to the ball without also imparting a translation force. Hence, the device is often restricted to three degrees of freedom—translations or rotations, but not both simultaneously. One such torque converter is the Spaceball 2003, marketed by Spaceball Technologies.

Magnetic Field Devices

A magnetic field device uses a source point, usually located on a pen-like wand, that emits a low-frequency magnetic field. The magnetic field activates a sensor mounted below a platform. An object to be digitized can be placed above the platform, or the wand can be used freehand in the space above the platform. Magnetic field devices capture the x, y, and z coordinates of a selected point and the directional attitude of the wand at that point. Hence, they have six degrees of freedom. Examples of magnetic field devices are the the 3SPACE Digitizer and the 3SPACE Isotrack, both marketed by Polhemus. The Isotrak comes with a sensor consisting of a small triad of electromagnetic coils enclosed in a plastic shell that senses a magnetic field as it moves through space. The Digitizer has a freestanding 40-inch model table that contains the sensor. Both can use either a source similar in shape to the sensor or a stylus. Both units also support a foot switch used to transmit data points from the electronics unit to the host computer. Metallic objects can cause distortions in the magnetic field. The Isotrak can recognize the source within a 30-inch sphere surrounding the sensor. The Digitizer supports models up to $18 \times 18 \times 10$ inches. Both systems can be extended to 60 inches of distance from source to sensor, but with reduced accuracy. Normal accuracy for either system is $0.1°$.

Acoustic Devices

Acoustic devices are similar in nature to magnetic field devices, but they rely on an ultrasonic signal sent from a base unit to the mobile unit to determine position. The 2D/6D mouse, created by Logitech, can function as a standard three-button mouse hooked to a personal computer and moved about on a desktop in 2D mode. It can also be used as a six-degree-of-freedom input device, reporting x position, y position, z position, pitch, yaw, and roll over a three-dimensional area in 6D mode. In addition to a control unit, the 2D/6D mouse consists of a mouse unit and a triangle unit. The mouse unit consists of a main body and a front module that houses three receiving microphones. The body is designed for holding the mouse in the air for 6D operations and on a table for 2D operations. In 6D mode the mouse has a tracking area or drawing volume that is a 2-foot cube with 200 dpi resolution, or a 7-foot cube with 10 dpi resolution for use as a head tracker.

Other Devices

The boom [LEVI91] is a device that has several joints and moves in a way similar to a human arm. Though designed to take the place of a head-mounted tracking system, it can be used as a 3D input device. Optical sensors on the joints provide high-precision feedback for the determination of the location and orientation in space of a pencil with buttons to direct selection, dragging, drawing, and other actions. The drawing volume is limited only by the length of the arm of the device and is considerably larger than the volume for either of the two previously described devices.

7.3 Application: An Interactive 3D Paint Program

In [BUTT88b] an interactive drawing program is described that was implemented on a field-sequential 60 Hz shutter system. In the attempts to implement an interactive stereoscopic cursor as a tool to define and manipulate stereoscopic objects, several interesting stereo phenomena were manifested.

The drawing modes included a point-marking mode, a continuous marking mode useful for polygon fill, and a rubber-band vector mode. When the fill color was the same as the background, the continuous fill mode was suitable for erasing objects as well. The cursor was a 2D bit-mapped image. The system provided interactive control over a wide range of parameters for stereo pair generation, color table specification, position sampling rate, cursor size, cursor velocity, and other variables.

The cursor was an effective drawing tool, as the stereo pair in Colorplate 7.1 suggests. Users will generally be able to perceive a one-pixel change in cursor parallax. If 10 inches of arc (0.0028°) is assumed to be normal human stereoscopic acuity [TYLE77], parallax changes as small as 0.05 mm can be detected at a viewing distance of 1 meter. This implies that resolutions an order of magnitude greater than 640×512 on a 340 mm $\times$ 250 mm screen would still allow the human visual system to recognize a one-pixel change in separation as a change in depth.

When the cursor was near another object, for example, within twenty to forty pixels (10–20 mm), the parallax of the object image could generally be matched exactly by the user, resulting in identical depths. If the cursor was far removed from the object—on opposite sides of the view volume, for example—relative parallax matching was less accurate, with errors in the range of five to ten pixels. When depth-matching speed is important, additional cues may improve this performance. We discuss this in a later section on stereo cursor types.

Stereo Pair Generation Methods

Stereo pairs were generated using the rotation model, which is discussed in Chapter 5. The addition of a wireframe outline of the view volume, called a *reference cube,* gives a powerful sense of three-dimensional space and volume. Colorplate 7.1 illustrates the reference cube. The rotation angle was 9° about a vertical axis located in the middle of the front face of the reference cube. The linear perspective projection was such that the ratio of the sizes of far and near faces of the cube was 0.67. Having the edges of the reference cube recede toward a vanishing point and having the cube faces decrease in size are significant depth cues.

Induced z Shift

The field-sequential method used in this system produced an interesting result: a change in apparent cursor depth caused by horizontal cursor motion. Movement to the right causes an apparent decrease in depth, and movement to the left causes an apparent increase, with correct depth returning as soon as motion ceases. This phenomenon is caused by a combination of the system's slowness, the image display sequence, the phosphor decay rates, and the user's short-term visual memory.

As an illustration, consider right cursor movement in five-pixel increments at an arbitrary depth resulting from a left-right cursor image parallax of ten pixels. A cursor pair is always shown with the left image first, followed by the right, followed by the next left-right pair, and so on. In Figure 7.1 the first left-right pair (L1-R1) is drawn ten pixels apart. Following the five-pixel move to the right, the next cursor of the following left-right pair (L2-R2) is drawn. Before the corresponding new right cursor is displayed, the old right cursor is still visible, and it and the new left cursor image form a stereo pair with a reduced parallax of five pixels, not the original ten.

When the new right cursor, R2, is displayed, the parallax will be ten pixels once again. As a result, during motion to the right, the average cursor image parallax alternates between five and ten pixels, giving an average parallax of $(10 + 5)/2 = 7.5$ pixels. This causes a decrease in average apparent depth solely as a result of horizontal motion. This situation is shown in Figure 7.2.

Erasing in Stereo

In normal 2D paint programs, part of an object can be removed by over-painting it with another object in the background color. In stereo, if the user wants only to partially erase the underlying object, the parallax at the edges

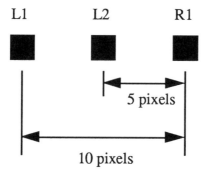

Figure 7.1
Left-right pair with ten-pixel separation.

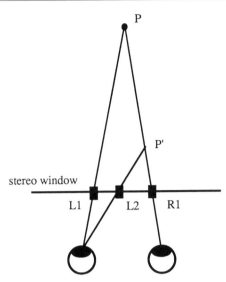

Figure 7.2
Left-right pair with average depth.

of the background-colored object may be discernible, and it may appear as a solid object at a different depth than the object being erased. Attempts to erase an object when the cursor is not exactly at that object's depth do not result in erasure, but rather produce a new "object" in the background color at the new depth. The three black dots within each of the two rectangles below the squares in Colorplate 7.2 clearly illustrate this problem.

It is obvious from Colorplate 7.2 that for stereoscopic 3D paint programs, a new requirement for hidden-surface elimination is that any partial erasing must be done only when the cursor z-value exactly matches that of the object to be erased. Hence, for hardware z-buffer algorithms, the user must be able to select a mode that overwrites only when cursor z equals buffer z. Without this feature, partial erasing requires very precise placement of the cursor in depth.

On the other hand, when constructing new objects in a random sequence, the user must be able to select a mode that overwrites when the cursor is in front of an underlying object, but does not overwrite when the cursor is behind an existing object. This mode would be the most common for painting operations. Without such a mode, other stereoscopic ambiguities result, as can be seen where the rectangles overlap in Colorplate 7.3. These rectangles were drawn in left-to-right order with z increasing for each new rectangle; thus, the right edges appear to belong to the deeper rectangle to the right, but they actually belong to the nearer rectangle on the left. Without this mode, the only way to paint unambiguous objects is to construct the objects in depth order starting with the deepest.

Rubber-Band Vectors

The rubber-band vector mode works well if vectors can be drawn within the refresh speed of the frame buffer, preventing buffer switching on the same retrace pulse used for the shutters, which would result in both eyes seeing the same image. With workstations or graphics boards that are "stereo ready," this is not a problem since quad buffering is supported. We discuss this in more detail later.

Often, as the vector length is increased, the stereo sensation remains, but with noticeable flicker or cross talk between left- and right-eye images. If the vector requires too much drawing time, proper synchronization may occur, but only on alternate signals. In this case each eye is presented with one of the opposite eye's images for every two of its own. Despite these limitations, effective drawing can be accomplished using the rubber-band vector mode since the flicker or cross-talk conditions are usually short in duration.

Polygon Fill Speed

The cursor must be able to be erased or redrawn within the refresh speed. The z-buffer speed and polygon fill speed thus limit its size and shape. This may preclude the capability to paint (or erase) wide swaths in the continuous marking mode as the cursor moves through the view volume.

Picket Fence Problem

The raster technology produces another perceptual problem that is most noticeable in the rubber-band vector mode. This is called the picket fence problem because as a vector that is oblique in z moves through the vertical, it is partitioned into a series of vertical pickets staggered in depth. A side view of the phenomenon is illustrated in Figure 7.3.

This phenomenon occurs only when the vertical segments of the diagonals in each left and right image are closely matched in length. This means that the vectors in each left and right image must be nearly mirror images of each other. The problem is most noticeable when a vector has approximately a 45° slant in z and is within 10–20° of the vertical in x and y. Beyond these vertical parameters in x and y, sufficient differences in left and right

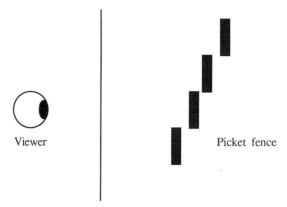

Viewer   Picket fence

Figure 7.3
Picket fence problem.

Plates

Several colorplates consist of three images representing two stereo pairs of the same scene. The leftmost image is a left-eye view, the center image is a right-eye view, and the rightmost image is another left-eye view. These images can be viewed without special devices by using a technique called free viewing (see Chapter 1). There are two forms of free viewing: parallel, or uncrossed, viewing and cross, or transverse, viewing. To parallel-view, use the two left images. Place a piece of paper with its edge between the two images and locate your eyes so that the left eye can see the left-eye image but not the right-eye image and vice versa. The two images should "fuse" in the center, and after a few seconds most viewers can see depth. You may have to cross your eyes slightly to form the stereo image. To cross-view the stereo pair, use the two right images. Stare at the line between the two images and gently cross your eyes so that a single image forms in the center. Again, after a few seconds most viewers will see the scene with depth. Some viewers may find parallel viewing easier than cross viewing, while others may find cross viewing easier. The techniques may require some practice, and those who are stereoblind may not be able to apply either technique.

Plate 1.1 Holographic Stereogram

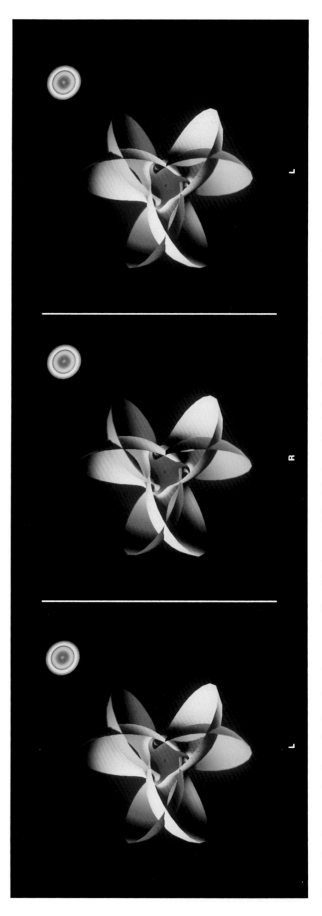

Plate 1.2 Stereo pair of a projective plane (courtesy of Lou Harrison and Paul Barham, North Carolina State University and Yates Fletcher, Sun Microsystems)

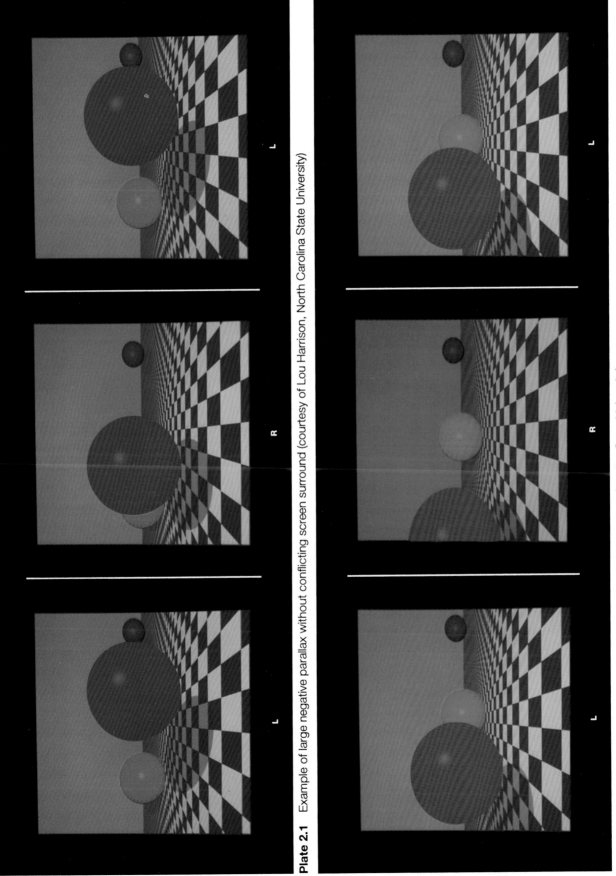

Plate 2.1 Example of large negative parallax without conflicting screen surround (courtesy of Lou Harrison, North Carolina State University)

Plate 2.2 Conflicting screen surround (courtesy of Lou Harrison, North Carolina State University)

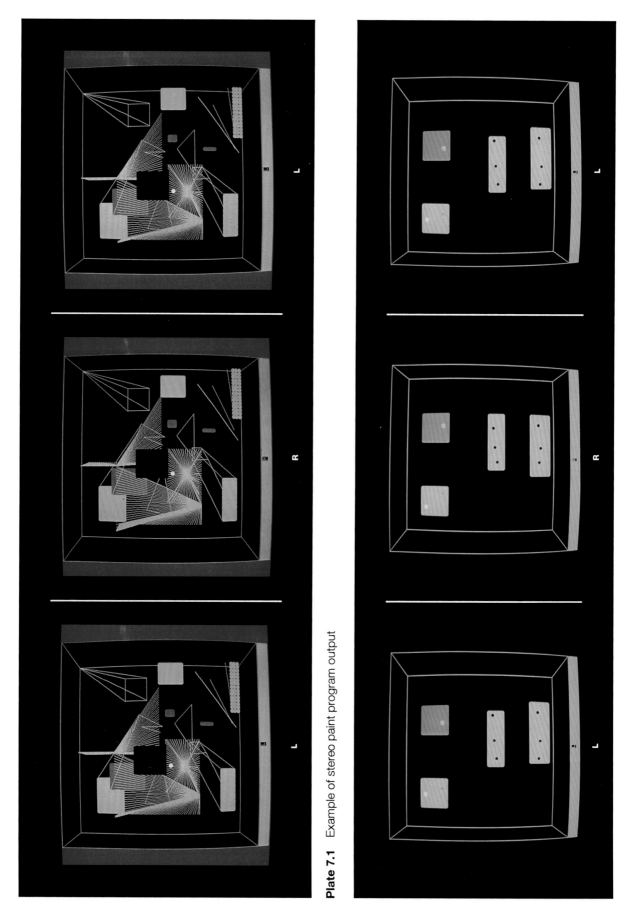

Plate 7.1 Example of stereo paint program output

Plate 7.2 Erasing in stereo

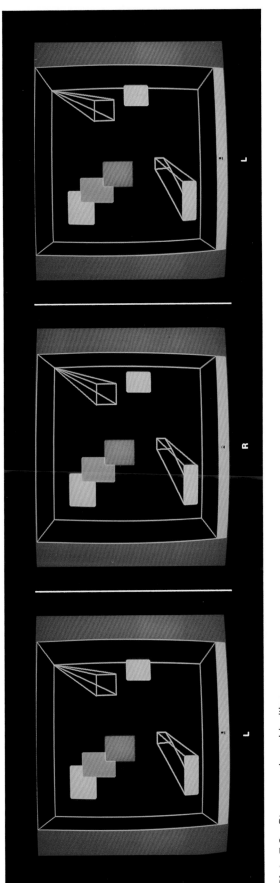

Plate 7.3 Stereoscopic ambiguities

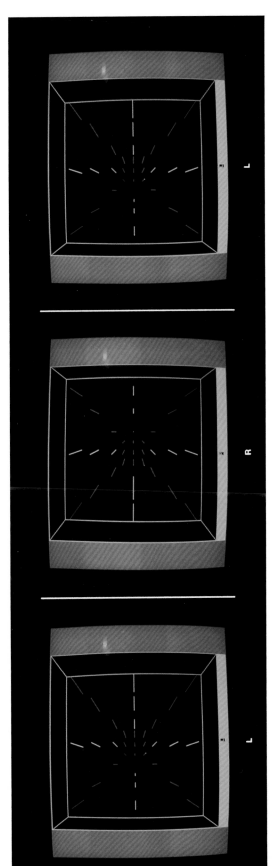

Plate 7.4 Double imaging

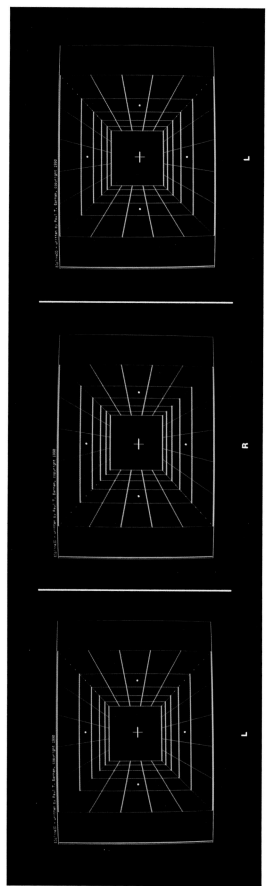

Plate 7.5 Grid and ghost points add depth perception

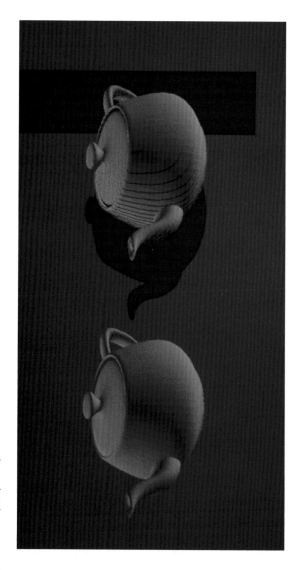

Plate 7.6 Stereo pair using pixel shifting, arranged for cross viewing

Plate 7.7 Stereo scene before color quantization

Plate 7.8 Stereo scene after color quantization

Plate 10.1 A chromostereoscopic simulation using a stereo pair (image created by Sam Shearman using AutoDesk 3D Studio, © 1993 Applied Physics Research, L.P.)

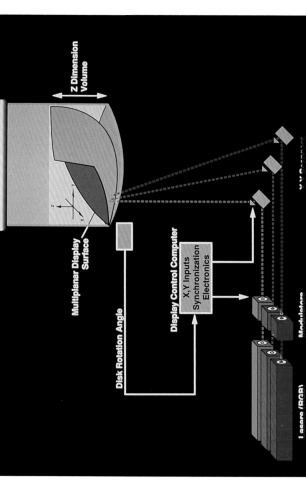

Plate 13.1 OmniView™ volume display system

Plate 11.1 BB&N varifocal mirror system

segment lengths exist so that the visual system tends to blend them and mask the problem.

The phenomenon is usually not bothersome since it is restricted to a fairly narrow range of parameters, and when it does appear, it does not degrade stereo perception significantly. Elimination of this problem will require higher resolution and antialiasing, both of which require increased hardware bandwidth.

Perceptual Zooming

Chapter 4 discussed the phenomenon called Emmert's law. If a 2D cursor does not have linear perspective, then as the depth of the cursor increases, cursor size will appear to *increase* rather than decrease. The term *size constancy* or *perceptual zooming* describes this phenomenon. Emmert's law states that as the convergence angle decreases (i.e., as objects become more distant), objects within about 1 meter from the viewer are perceived as increasing rather than decreasing in size. Size constancy arises from the fact that as an object moves away and appears to grow larger in accordance with this law, its actual retinal image decreases in size, resulting in a roughly constant perceived size within this 1 meter range [JULE71].

However, even though stereoscopic objects appear to move away from the viewer when the convergence angle decreases, they are in reality still fixed at the distance of the monitor from the viewer; thus, there is no decrease in retinal image size to compensate for the perceived increase in size—hence the term *perceptual zooming*. To avoid this phenomenon, if the user is implementing a 2D (or 3D) cursor, the size should be determined by a perspective projection. It may be more efficient to do a table lookup of cursor size based on parallax if the transformation speed of the frame buffer does not permit near real-time rendering of the scene.

Double Images

A more annoying and distracting phenomenon is the problem of double images. This is most pronounced for long, slender objects oriented along the z-axis and is perceived by the viewer as the inability to fuse the left- and right-eye images throughout the object's length. Colorplate 7.4 illustrates this case.

If the eyes are focused on either end of a long rod, double images are perceived at the other end. A further difficulty occurs when there is no point object or feature to maintain the viewer's visual attention, so that the point of convergence is unstable and the eyes tend to wander back and forth along the rod. Three factors tend to cause this phenomenon. First, the depth of field in stereo images is essentially infinite. In a real scene, a similarly oriented rodlike object does not exhibit this problem to the same extent because of the limited depth of field of the human visual system. This limited depth of field causes the extreme ends of a real object, where parallax differences are maximized, to appear blurred, possibly making them less distracting. The second factor is based on the fact that the viewer must learn to uncouple accommodation from convergence in stereo images, which introduces another subtle psychological distraction. Third, in a real scene, a viewer can move his or her head to generate motion parallax. This disambiguates confusing depth information and helps to reduce the double-image problem. Achieving motion parallax is obviously not possible in a stereo pair system without head tracking. Decreasing the maximum depth of the view volume or increasing the perspective ratio in the stereo image seems to reduce the double-imaging phenomenon. Depth-shading techniques may also decrease the effect. Also note that lines that recede from the viewer parallel to the z-axis and originate on the x-axis cannot be seen.

An Example of the Interposition Depth Cue

The stereo pair in Colorplate 7.3 illustrates how powerful the interposition depth cue can be; indeed it may appear to override and contradict other depth cues in an image. In the upper left corner, most viewers will at first see the green rectangle in front since it appears to obscure the one behind it. Once the viewer fuses the stereo pair, it will be clear that the green rectangle has greater depth than the other two. Some viewers never perceive the depth and insist that the green rectangle is in front, although they may see the stereo in other parts of the image.

7.4 Application: Interactive Manipulation of B-Splines

Interactive input techniques for 3D computer graphic systems have been widely researched for the past two decades. Most of this research was conducted using some type of 2D input device to manipulate either a 2D or a 3D cursor in a 3D scene ([LEVK84], [BIER86], [NIEL86], [BUTT88a]). Unfortunately, the mapping between the 2D input device control and the 3D cursor movement is not intuitive for the user.

With the introduction of six-degree-of-freedom input devices in the early 1980s, researchers were able to provide the user a one-to-one correspondence between the device control movements and the cursor's movements. Since the 3D scene was usually rendered on a two-dimensional screen, researchers found that the user was still unable to manipulate and position the cursor in an intuitive manner due to the lack of depth perception [BADL86]. However, Lipscomb [LIPS79] found that users can learn to position rapidly in three dimensions using a monoscopic display by positioning first in 2D with a view parallel to the xy plane, and then rotating to a view parallel with the yz plane and positioning horizontally across the screen, along the former z-axis. Biochemists using this method constructed several molecules of several thousand atoms each about as quickly as those who used stereo. This manipulation technique is called "motion decomposition" [KILP76].

Stereo systems are now being used in conjunction with six-degree-of-freedom input devices to yield true three-dimensional input and output ([SCHM83], [WALD86], [BEAT87], [BEAT88], [BURT88]). Interfaces for these systems are now important areas of research ([BRID87], [MOSH88], [WIXS88], [WIXS90]). An early such stereo cursor system developed by Richard DeHoff and Peter Hildebrandt of Tektronix [DEHO89] had a pointer symbol tethered to a reference symbol on screen, so that the cursor had the effect of rubber-banding out of the center of the view volume. The reference symbol was defined as a "+" and was attached by a straight line to an "×," which was the pointer symbol. The reference symbol had zero parallax, and the pointer symbol had parallax proportional to its position in the view volume. The parallax of the tether line varied from zero, at the reference end, to its maximum, at the pointer end. One of the interface issues not considered in this system is the type of stereo cursor to use for particular applications. For example, standard 2D cursors are acceptable for text manipulation, but cursors that communicate depth in a stereo environment must be chosen for the particular application.

Barham and McAllister [BARH91] described an interactive stereo system to allow a user to construct, draw, and modify B-spline space curves. They investigated several different cursor types to determine which were suited for this application. They also examined the characteristics that made particular cursors useful [HODG87]. Their system used a Spatial Systems Spaceball to control the stereo cursor while a mouse was used to manipulate a 2D cursor for menu selection. The toggling back and forth between the mouse and the Spaceball was annoying since the user was required to switch attention from the screen to the particular input device and then back to the screen. When possible, stereo systems should be controlled by a single input device. This, however, can produce other problems, as indicated in Section 7.5. We describe briefly the capabilities of the Barham system to motivate the choice of cursor types that were discussed in the project.

System Overview

The system executes on a Sun 3/260 color workstation with a TAAC-1 application accelerator (TAAC). The TAAC is a parallel graphics accelerator programmable in C. The program allows the user to create, select, delete, and modify B-spline space curves interactively in stereo. To create a curve, create mode is entered, and a small circle is attached to the drawing point of the cursor. The user can then move the cursor with either linear, planar, or spatial freedom within the view volume defining control points by pressing the select button on the Spaceball. The control polygon and resulting curve are interactively drawn up to either the last control point defined or to the current cursor location, depending on whether rubber-band mode is off or on, respectively. The user has the capability to modify the B-spline order and specify whether a B-spline is periodic uniform or open uniform. There are also options to toggle drawing a control polygon around the spline and to manually enter individual points. To complete a curve, the user double-clicks the select button at the last control point. The curve remains selected, and the program defaults to select mode.

The user selects a curve by positioning the cursor on the curve and pressing the select button. If the cursor is sufficiently close to the curve, the curve becomes active and small circles are placed at the control points. If the cursor is not sufficiently close, no curve is selected. In world coordinates, the cursor must be no more than 25 Euclidean distance units away from a computed point on the curve to be sufficiently close. The view volume is a 1,024-unit cube in world coordinates. This closeness amount can be varied from one-half to two times its default value interactively via a slider. An additional feature is curve and point gravity. If the gravity feature is enabled, then pressing the select button causes the cursor to attach either to the nearest control point or to the nearest curve, depending on whether or not a curve was previously selected. Once a curve is selected, it becomes the active curve, and future modifications affect only this curve. An active curve is deleted by choosing the delete button option. This erases the curve from the screen and leaves the program in select mode with no curve selected.

The user can modify curve characteristics such as the color of the curve, the color of a curve's control polygon, the knot vector, and the order of the curve. The control polygon can be turned on or off. The curve can be opened or closed. To adjust a control point on the active curve, the cursor must be placed on the control point and the select button pressed. This attaches the control point to the cursor, changes the control point color to that of the cursor, and allows the user to interactively move the point while observing the corresponding changes to the curve. To release the control point, the select button is pressed at the desired location.

For computing the stereo pairs, Love [LOVE90] has proposed a method involving a shear operation instead of a translation that will easily allow the stereo window to be placed anywhere in the scene. With this system, the user can adjust the eye-to-screen distance in inches and the location of the stereo window in world coordinates within the scene. The system also has the added option of moving the center of projection nearer to or further from the viewplane in order to control linear perspective. The interocular distance, e, is fixed at 2.5 inches. Users who are unable to fuse images computed with an interocular distance of this size can indirectly reduce the interocular distance by increasing the program's eye-to-screen distance while actually remaining a constant distance from the screen. The result is satisfactory stereo viewing with total user control.

Hardware

The TAAC has built-in routines to facilitate stereo imaging [SUN88]. These routines will automatically switch between displaying the red and green channels at refresh rates so that the right-eye image can be stored in one channel while the left-eye image is stored in the other. The stereo modulator and shutter synchronize the viewer's eyes with the appropriate images shown on the screen. Unlike the paint application described earlier, this system makes use of quad buffering of images. The current image being displayed is stored and refreshed from one bank of the frame buffer while the next image is being computed in the other bank of the frame buffer. For each image in time, one eye view is placed in the red channel, and the other is placed in the green channel of the drawing bank. The TAAC uses the vertical sync signal as an interrupt vector to switch the display from the left-eye view in the current display bank to the right-eye view and vice versa, so the alternation of the stereo pairs is totally independent of the drawing time, and phase reversals cannot occur. Also, each eye view is always refreshed at a frequency of 30 Hz. When drawing is completed, the drawing and display banks are switched. The effect of the quad buffering, or double-double buffering, is that the user never sees a new scene until it is completely rendered. Since the red and green channels are used for the left and right images, the two remaining channels, the blue and alpha channels, can be jointly used as a 16-bit z-buffer for hidden-surface elimination.

Cursor Types

Several 2D and 3D cursor types have been described in previous research (e.g., [BIER86], [NIEL86], [BUTT88a], [BEAT88]). The 2D and 3D cursor shapes are shown in Figure 7.4. The full-space crosshair and jack are shown bounded by the the view volume in Figure 7.5 and Figure 7.6, respectively.

The point on the cursor that is used to select a given point or create a new point is referred to as the hotpoint. The hotpoint of the circle, square, sphere, or cube is at the center of the object. The hotpoint of the triangle, pyramid, or arrowhead is at the apex of the object. The hotpoint of the crosshair, full-space crosshair, tri-axis, jack, or full-space jack is at the intersection of the vector components of the object.

The cursors are all polygonal or vector objects. The polygonal objects can be wireframe, flat, or Gouraud-shaded. The cursors are interactively z-buffered, perspectively projected, and redrawn after each user operation, as are all objects in the scene.

Figure 7.4
2D and 3D
cursors.

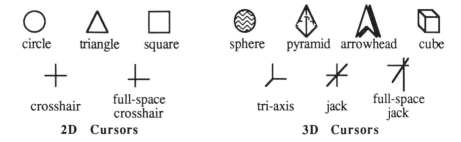

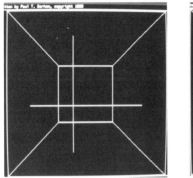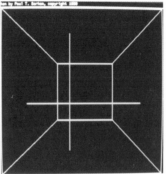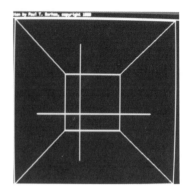

Figure 7.5 Full-space crosshair.

Also, as shown in Figures 7.5 and 7.6, a wireframe cube is always drawn surrounding the view volume since this has been shown to enhance depth perception ([BUTT88a], [BEAT87], [BUTT88b]). Colorplate 7.5 shows how a grid can be overlaid on the cube to present more texture in the scene and increase the depth sensation. The hotpoint of the cursor is confined to lie within the reference cube.

The hotpoint can be projected onto the sidewalls of the reference cube as a cue to the cursor location in the view volume. These projected points are referred to as the cursor's ghost points. The use of ghost points in Colorplate 7.5, in conjunction with the grid reference cube, can give an indication of absolute position of the cursor within the view volume.

Cursor Selection and Placement

One concern in stereo interface design is the accuracy with which the user can place a cursor's hotpoint when selecting a control point. Several subjects participated in an experiment to select and move all the control points for a test curve. The experiment was performed in stereo with the

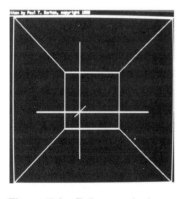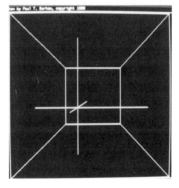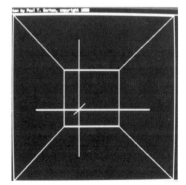

Figure 7.6 Full-space jack.

stereo window at the front of the scene. The subjects were told to strive for accuracy and not to worry about time constraints. Each subject was allowed to choose the color of the reference cube, the cursor, and the curve used in the experiment. The subject also controlled the size of the cursor. However, the user was not allowed to rotate the cursor. The subjects were allowed to move a stereo cursor around using the Spaceball prior to the actual experiment in order to gain a feel for the device. Most participants found the Spaceball overly sensitive and chose to decrease its sensitivity. Proficiency in the use of the Spaceball requires repeated exposure. During the experiment, the number of errors in control point selection were counted, and the user was asked to give a subjective ranking of the cursor on a scale of 1 (worst) to 5 (best).

The subjects used stereopsis for rapid initial placement of the cursor near the control point. Once the cursor was near the control point, interposition was used almost exclusively for fine placement of the cursor on the control point. Thus, the cursors that were most effective are the ones that give good disparity cues and ones that are easy to use for interposition. The additive nature of depth cues seemed to apply, since cursors that provided only one of the cues were consistently ranked lower than those providing both.

A shaded circle is effective in control point selection. The size chosen for the circle was close to the size of the circles representing the control points. The color for the circle was one that showed good contrast to that of the control points. The flatness of the circle enabled interposition to be determined with one-pixel effectiveness in the z dimension; thus, no errors were made. The size and shape of the circle provided effective parallax and enabled the circle to be positioned accurately even without the use of interposition.

The crosshair and jack in both regular and full-space modes are also excellent cursors but require more effort to represent and render. Subjects reported an extra sense of depth, but some disliked the z-axis line on the full-space jack since it can be hard to simultaneously fuse both ends of the line. Also, the two lines of the crosshair cursor do not appear to be at the same depth. Adding a small circle at the origin of the crosshair or jack helps to bring the lines to the same depth and provides a stronger interposition cue.

The triangle and pyramid were ranked as average, with size having little significance unless the object was scaled large enough to obscure other objects in the scene. The pyramid provided interposition cues comparable to those of the triangle since the thickness of the pyramid near the apex is small. When the square was scaled to the approximate size of the control points, it was hard to distinguish from the circle and was accordingly highly ranked. The sphere and cube cause difficulty in conjunction with interposition. The tri-axis was not a good cursor for this application. The gravity feature is extremely convenient if it can be toggled on and off with ease.

Drawing New Curves

Another stereo interface design concern is the user's ability to create new curves. The same subjects were asked to perform a second experiment in which the task was to reproduce an existing curve as accurately as possible. The new curve could be translated only in x, and the y and z dimensions were to match the original curve as closely as possible. The user was not allowed to use interposition from the original curve; thus, no overlap of the two curves was permitted. Both rubber-band and regular modes were tested. The subject was asked to draw from experience in the previous experiment and to choose the cursor he or she thought best for drawing. Some form of the jack was usually chosen. The ratio of inter–control point distances was compared between the new curve and the original. Absolute errors in y and z were also examined between the two curves. No obvious benefits resulted from rubber-banding, with the user being more accurate about half the time. This lack of improvement may be attributed to the test curve being a difficult curve to reproduce due to the little change in x and y and the large changes in z. However, the user always subjectively ranked the rubber-banding mode as superior. The rubber-banding gives a point of reference, a sense of direction, and an estimation of distance to the user.

Since interposition was not allowed, other cues became important for the drawing task. Full-space cursors were frequently chosen due to their ability to deliver position within the view volume. When the full-space cursors were not chosen, users always chose to use the ghost points for the extra sense of location they provided. The grid was always preferred on the reference cube for its sense of scale and the extra sense of texture reference. Visual enhancements play an important role in stereoscopic tracking tasks [KIM87].

Summary of Results

In summary, cursors that provide good depth cues and an indication of location within the view volume are superior for applications such as this one. Full-space jacks and crosshairs give a good sense of location within the view volume. Regular jacks and crosshairs combined with the use of ghost points are also effective indicators of position. Using a small circle cursor at the origin of a jack or crosshair yields sufficient parallax information and excellent interposition cues due to its flatness. In drawing with the cursor, rubber-banding is important. Depending on the application, the rubber band should start from a logical and useful point that will not interfere with the rest of the scene. For selecting points with the cursor, a gravity feature should be an option. A cube with a grid texture surrounding the view volume supplies a needed reference.

Perceptual Problems Encountered

The first perceptual problem encountered has already been mentioned. The horizontal and vertical lines of the full-space crosshair and jack cursors do not appear to be at the same depth. This perceptual phenomenon may be caused by the fact that the hotpoint is the center of attention, and the horizontal line of the full-space crosshair does not present any parallax near this point. The parallax for the horizontal is in the peripheral vision and does not yield as strong an effect as foveal attention [ANDE90]. Placing a small circle at the origin of the crosshair appears to bring the depth of the horizontal and vertical lines together. This may be a grouping effect due to the identical color, parallax, and proximity of the objects [NAKA89]. Combining the crosshair and jack, especially full-space models, with the circle is a perceptually strong combination yielding good depth and interposition cues. A second way to bring the depth of the crosshair lines together involves drawing the lines of the crosshair at oblique angles rather than parallel to the x and y axes. This would allow parallax information to be present all along the crosshair, and especially at the origin. Another factor that may influence the crosshair line separation is line width. The built-in line-drawing algorithm does not take into account the aspect ratio or addressability of the monitor. This shortcoming causes horizontal lines to be wider than vertical lines. Since relative size is a depth cue, the differing line widths may contribute to the perceived depths of the lines of the crosshair [REIN90]. Antialiasing should be applied to all lines to create uniform line widths and to prevent differing amounts of aliasing for the same line in the different eye views.

The second perceptual problem is that when the head is moved, the scene is perceived to change. This problem is inherent in the geometry used to compute the stereo views. Hodges, Love, and others ([HODG92], [LOVE90], [WALD86]) have discussed the phenomenon of image scaling that occurs when the head is moved closer to or further away from the screen. Since the application has parameters to control the eye-to-screen distance, this problem can be overcome to some extent. However, when the head is moved side to side, the scene appears to rotate about some y-axis depending on the location of the stereo window. Points at the level of the stereo window appear not to move as the head is moved side to side. However, points behind the stereo window move opposite to head movements, and points in front of the stereo window move with head movements. Examples of this phenomenon are shown in Figure 7.7 for different locations of the stereo window in the same scene. Gogel discusses the psychological and perceptual aspects of this phenomenon in his paper [GOGE90]. To solve these geometry-related problems, a head-tracking device could be used to calculate current viewing parameters. A more feasible solution is to use a head-mounted stereoscopic display system.

When possible, the view volume should be restricted to the region that is viewable by both eyes, or the scene should be clipped to this region. Since the perspective projection to each eye results in different truncated pyramid shapes, it is possible to have an object that the left eye views and the right eye does not, and vice versa. The possibility exists, in the extreme

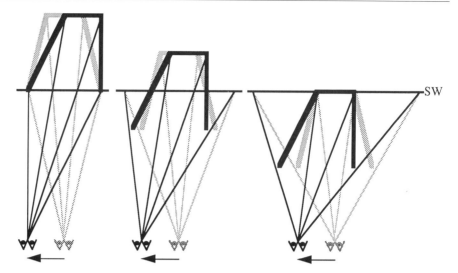

Figure 7.7 Image distortion resulting from head movement to the left with the stereo window at the front, middle, and rear of the scene.

case, that each eye will see an object the other eye does not, and that these objects are so similar that they are merged as one. As shown in Figure 7.8, the objects will fuse in front of the stereo window. Objects placed in these nonoverlapping areas have no depth information. These objects will be distracting, cause eye strain, and make the scene difficult to fuse.

7.5 Menuing Interfaces in a Windowing Environment

In [CARV91] an interactive drawing program was implemented for an autostereoscopic DTI display, which is described in Chapter 9. The display permits the simultaneous presentation of stereo pairs as opposed to the field-sequential method. The system runs on a Macintosh II and is a stereo implementation similar to the well-known MacDraw object-oriented graphics program. The Macintosh interface was followed as closely as possible, as was the MacDraw menu system, with some modifications to facilitate the stereo nature of the program. The effort to preserve the MacDraw interface raised several issues about the design of a stereo interface and the features that should be present in an operating system to facilitate the development of stereo applications.

Menu Design

It is difficult for most stereo users to fuse abrupt changes in parallax. For example, if a cursor makes discontinuous changes in depth quickly, the visual system typically needs some time to adjust to the different position

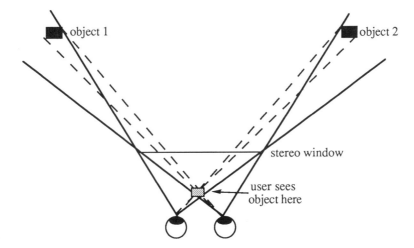

Figure 7.8
Pseudomerging
of partially
clipped objects.

of the cursor and to fuse the left- and right-eye images. A rapid change in negative parallax normally takes a few seconds to fuse.

Stereo software systems have traditionally been designed to produce output in one or more stereo windows. These stereo windows are part of the 2D windowing system of the workstation. The left- and right-eye views of the windowing system menus and window boundaries are identical and have zero parallax; hence, they will appear in the plane of the display surface. If the user wishes to access services from the 2D menus or windows of the 2D windowing system while working in a stereo window, the visual system has to adjust to the discontinuous change in parallax caused by the transition from viewing an object in stereo to viewing text with zero parallax. This can be very disconcerting and annoying, and it can cause fatigue.

An obvious solution is to allow the operator to signal that an abbreviated (tear-off) menu should appear at the same approximate depth as the cursor. The user can then select actions from the menu while continuing to focus at the same depth. This technique has been used in virtual-reality applications.

Text

The treatment of text in a stereo environment can become complicated and increase rendering time considerably, depending on the generality required in the application. If text is to obey linear perspective under translation, the operating system utilities must be able to produce text of arbitrary font size. Often text is represented by bit maps that can be scaled from a finite set of installed fonts, which can produce unsatisfactory text shape and considerable aliasing.

To retain maximum image resolution, an outline font handler such as Adobe Type Manager in the Macintosh environment can be used to generate any fractional text height from the outline font representations. When a text object is moved in depth, its projected height is updated to maintain its consistent virtual height (string width is scaled automatically by using the appropriate font size). Normally, the font size after translation in depth would be stored in a table as a function of its size at zero parallax (at the depth of the stereo window) and its current depth.

Text handlers normally assume that the text is to be generated in a plane parallel to the viewing screen. If rotational transformations are also to be applied to text, then text handling becomes much more complicated. The capability of modifying the outline definition of a font to handle rotation and linear perspective is not currently available in existing text handlers such as Postscript. Outline fonts are often described using Bezier or Hermite splines. Neither of these splines is invariant under projection [ROGE90]. This implies that the entire curve must be drawn, rotated in 3D, and then projected. Nonuniform rational B-splines (NURBS) are invariant under projection, which means that the rotation and projection need be applied only to the control points and the resulting curve drawn in 2D. An alternative is to place a bounding rectangle around text and apply the transformations to the rectangle. The equivalent 2D transformation (after projection) can then be applied to the text in the rectangle. Bit-mapped fonts can suffer from considerable aliasing with this approach. Obviously, these problems do not arise if text is restricted to be parallel to the stereo window.

Hidden-surface elimination can also become complicated if text is treated as two-dimensional isolated characters. An alternative is to treat characters as solid objects and send each character through the perspective transformation and hidden-surface elimination. This has the disadvantage of increasing the complexity of the scene dramatically and slowing rendering time.

7.6 Stereo Algorithms

Rendering of left- and right-eye views need not require twice the time of rendering a single frame. Simplifications can be exploited to reduce the required rendering time in certain cases. We will discuss two such algorithms: z-shear, proposed by Love [LOVE90], which is an extension of the shear technique described in Section 5.4, and ray tracing, described by Adelson and Hodges [ADELn.d.].

z-shear Technique

A potentially severe drawback to the use of stereo, particularly panoramic stereo, is the increased amount of computations required due to the many perspective images. This increased load could significantly lengthen the time required to generate the images to be displayed. Although this is disturbing for hardcopy, for a system attempting to operate in near real time it is prohibitive. For the display to be useful as an interactive output device, a user must be able to manipulate the image being displayed without experiencing significant lag time.

When a user manipulates an image on an interactive display, the primary concern is update speed rather than image quality. Although maintaining high quality is clearly desirable, if significant speed increases can be obtained, some degradation is acceptable. The varifocal mirror system developed at the University of North Carolina–Chapel Hill [FUCH82] could display a point cloud with maximum quality while the image was stationary. During manipulations, however, it was not possible to update the entire point cloud quickly enough, so only a fraction of the points were displayed. This reduced the load on the system to enable it to keep up with user requests. The image displayed during manipulations was certainly lower in quality but was adequate for the user to orient the object as desired. Once the image became stationary, there was no longer any problem with lag time, and the full-quality image was again displayed.

A similar acceleration method can be employed with panoramic stereo. When computations exceed what can be performed in the time available, it is possible to sacrifice quality in some of the perspective images being computed so that, unlike the case for the varifocal mirror, there will always be at least one image that is computed at maximum quality. There is a great deal of similarity among these different images, and this can be exploited [HODG92]. Once a single image has been computed, other images can be quickly approximated. The following strategy uses linear interpolation to shift pixel values to approximate different perspective images and is an extension of the shear stereo transformation presented by Love [LOVE90].

A shear is a linear transformation that has the effect of tilting lines. For example, italics can be formed from regular letters by applying a shear transformation. If the x-coordinate of a point is sheared by z, the result is that x is incremented by some scalar multiple of z:

$$[x \quad y \quad z \quad 1] \begin{bmatrix} 1 & 0 & 0 & 0 \\ 0 & 1 & 0 & 0 \\ S & 0 & 1 & 0 \\ 0 & 0 & 0 & 1 \end{bmatrix} = \begin{bmatrix} x + Sz \\ y \\ z \\ 1 \end{bmatrix}$$

To differentiate stereo images, different shear values appropriate to each eye position are used to produce horizontal parallax while both the y and z coordinates are left unchanged. Parallax between corresponding image points from two different perspective views is a function of the z-coordinate. Points on the plane $z = 0$ at the time the shear is applied will have zero parallax. When displayed, they will appear at the depth of the display panel. The stereo window, and therefore the display panel, can be located at any depth in object space simply by translating from the desired depth to the plane $z = 0$ before shearing and then translating back. When projected, points at the desired depth will have zero parallax. The sequence can be represented as follows:

$$
\begin{bmatrix} 1 & 0 & 0 & 0 \\ 0 & 1 & 0 & 0 \\ 0 & 0 & 1 & 0 \\ 0 & 0 & -z_w & 1 \end{bmatrix}
\begin{bmatrix} 1 & 0 & 0 & 0 \\ 0 & 1 & 0 & 0 \\ S & 0 & 1 & 0 \\ 0 & 0 & 0 & 1 \end{bmatrix}
\begin{bmatrix} 1 & 0 & 0 & 0 \\ 0 & 1 & 0 & 0 \\ 0 & 0 & 1 & 0 \\ 0 & 0 & z_w & 1 \end{bmatrix}
\begin{bmatrix} 1 & 0 & 0 & 0 \\ 0 & 1 & 0 & 0 \\ 0 & 0 & 1 & 1 \\ 0 & 0 & 0 & 1 \end{bmatrix}
$$

$$
\quad\ \text{Translate} \qquad\quad \text{Shear} \qquad\quad \text{Translate} \qquad\quad \text{Project}
$$

Let d be the desired distance to the stereo window and S be the shear required to move the center of projection onto the z-axis. The consecutive eyepoints are separated by a distance e, the centermost eyepoint is on the z-axis, and the various centers of projection are located at $(ie, 0, -1)$ where i is an integer in $[0, \pm 1, \pm 2, \ldots]$. Then

$$
S = \frac{ie}{d} \tag{7.1}
$$

Composing these into a single transformation, we have

$$
\begin{bmatrix} x & y & z & 1 \end{bmatrix}
\begin{bmatrix} 1 & 0 & 0 & 0 \\ 0 & 1 & 0 & 0 \\ S & 0 & 1 & 1 \\ T & 0 & 0 & 1 \end{bmatrix}
=
\begin{bmatrix} (x + Sz + T) \\ y \\ z \\ (z + 1) \end{bmatrix}
$$

where $T = -Sz_w$, and

$$
\begin{bmatrix} x' & y' & z' & 1 \end{bmatrix} = \begin{bmatrix} \dfrac{(x + Sz + T)}{z + 1} & \dfrac{y}{z + 1} & \dfrac{z}{z + 1} & \dfrac{z + 1}{z + 1} \end{bmatrix} \tag{7.2}
$$

Being off-axis affects only x', the x-coordinate of the projected point, which can be rewritten as follows:

$$
x' = \frac{(x + Sz + T)}{z + 1} = \frac{x}{z + 1} + \frac{Sz + T}{z + 1} = \frac{x}{z + 1} + \Delta x' \tag{7.3}
$$

$$
\Delta x' = \left(\frac{Sz}{z + 1} \right) + \left(\frac{T}{z + 1} \right) = S \left(\frac{z}{z + 1} \right) + T \left(\frac{1}{z + 1} \right)
$$

$$
= S \left(\frac{z}{z + 1} \right) + T \left(\frac{z + 1 - z}{z + 1} \right)
$$

$$
= S \left(\frac{z}{z + 1} \right) + T \left(1 - \frac{z}{z + 1} \right) \tag{7.4}
$$

$$
\Delta x' = Sz' + T(1 - z') \tag{7.5}
$$

$\Delta x'$ is the change to the projected x-coordinate of a point when the center of projection is moved off the z-axis. Since the eyepoints are equally spaced, a single value for $\Delta x'$ can be determined for the eyepoint at $(e, 0, -1)$, and the value of $\Delta x'$ will be a constant multiple of this for all other eyepoints:

$$\Delta x_i' = i \Delta x' \tag{7.6}$$

and,

$$x_i' = x' + \Delta x_i' \tag{7.7}$$

Equation (7.7) is a very significant feature and can be applied to any point for which z' is available. The vertices of polygons can easily be projected for one eye position and that polygon rendered. Then, using the projected depth, z', each vertex can be shifted and the polygon rendered for all other eyepoints. This means, however, that the process of rendering each polygon must be repeated for each eyepoint. On the other hand, if hidden surfaces are removed by using a z-buffer, then the projected z-coordinate is available for every pixel on the display. This means that it is possible to render a single image for one eyepoint and produce images for all other eyepoints by simply shifting pixels (see Figure 7.9).

There is a problem with hidden surfaces if pixels are being shifted. It will almost surely happen that there will be surfaces that should be visible in a later image that were not visible in the first. Since these images were not present on the screen, shifting pixels around cannot produce them. Instead, gaps will appear in the image wherever new surfaces become exposed (see Colorplate 7.6). The extent to which this is a problem depends on the image, but it is significantly less troublesome than the standard hidden-surface problem. In the later case, it is possible to introduce objects into a scene that should not be visible. Incorrect features added to an image are hard to overlook. In the present case, however, errors occur in the form of gaps rather than additions and are much less problematic. After all, a

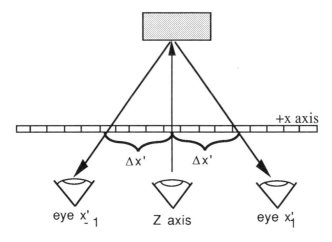

Figure 7.9
Calculating new eyepoints.

parallax barrier (described by Love in Chapter 3) introduces a tremendous number of vertical gaps in the image it presents, and its functionality is dependent on the ability of a viewer's brain to ignore omissions.

The pixel-shifting strategy should prove very beneficial in applications where speed is critical. For interactive manipulations, one of the stereo images can be computed and the others interpolated. Alternating which eye sees the interpolated images will help prevent gaps from becoming a problem. For higher quality, any desired number of views could be calculated exactly and intermediate images produced by interpolation.

Speedup for Ray-Traced Stereo Images

A variation of the method outlined in this section was first applied to speed up image generation for the frames of an animation sequence [BADT88]. Later it was adapted for generating simultaneous stereoscopic left-eye and right-eye images by Adelson and Hodges [ADELn.d.]. Since ray-tracing two separate images for the eyes doubles the work involved in rendering stereoscopic images, any savings that can be obtained proves useful. In this method, a technique is proposed in which the right-eye image is inferred from the ray-traced left-eye image. The color of a pixel is determined using only the initial intersection ray and a shadow ray; it does not allow for reflection and transparency rays [DEVA 91].

In stereoscopic images, a left-handed coordinate system is assumed with the viewplane and the stereo window located at $z = 0$ and the two centers of projection at $(-e/2, 0, d)$ and $(e/2, 0, d)$ for the left eye and right eye, respectively, where e is interocular separation. A point P projected for the left eye has viewplane coordinates

$$x_{sl} = \frac{x_p d - z_p(e/2)}{z_p + d} \qquad (7.8)$$

and

$$y_{sl} = \frac{y_p d}{z_p + d} \qquad (7.9)$$

The same point projected to the right eye has coordinates

$$x_{sr} = \frac{x_p d + z_p(e/2)}{z_p + d} \qquad (7.10)$$

and

$$y_{sr} = \frac{y_p d}{z_p + d} \qquad (7.11)$$

There is no keystoning or vertical parallax involved since the y-coordinates y_{sl} and y_{sr} are the same. A relation between the x-coordinates of the two views can be written as follows:

$$x_{sr} = x_{sl} + e\frac{z_p}{z_p + d} \tag{7.12}$$

Since the color of an object depends only on the position of the object and light sources, and not the viewing position, the same color that is assigned for the left eye can be given to the corresponding right-eye pixel. However, since specular highlights are dependent on the position of the viewer, the following modification must be applied to get the highlights for the right-eye view. The modification starts with Phong's shading model, where the intensity of specularly reflected light is given by

$$I_{sr} = SF[N \cdot (I + L)]^{\partial} \tag{7.13}$$

where

S	is the point light source intensity
F	is the spectral reflection coefficient
K	is the reflection coefficient of the surface
∂	is a "glossiness" factor
$N, I,$ and L	are normalized vectors representing the normal and the vectors to the eye and the light source, respectively.

From the Phong highlighting term, the intensity for the left eye can be written as

$$I_{srl} = SF[N \cdot (I_l + L)]^{\partial} \tag{7.14}$$

and the intensity for the right eye as

$$I_{srr} = SF[N \cdot (I_r + L)]^{\partial} \tag{7.15}$$

where I_{srl} and I_{srr} are the specular reflection intensities for the left eye and right eye, respectively, and I_l and I_r are the incident ray vectors to the left eye and right eye, respectively. Since $I_r - I_l = e$, I_{srr} may be written as

$$I_{srr} = SF[N \cdot (e + I_l + L)]^{\partial} \tag{7.16}$$

Equations 7.12, for x_{sr}, and 7.16, for I_{srr}, together provide the necessary information to simultaneously generate the right-eye image as the left one is being generated. This procedure is called *reprojection*.

Some problems occur with the reprojection procedure. First, it is possible to have more than one object project to the same pixel in the right-eye view. This can happen if an object intersects the reprojection vector back to the right eye. As an example, consider a ray from the right eye through pixel P1 at $(x_1, y, 0)$, as shown in Figure 7.10. Consider another ray, from the left eye through pixel P2 at $(x_2, y, 0)$. The two rays will intersect when the x values of the rays coincide. Using the x component of the ray equation,

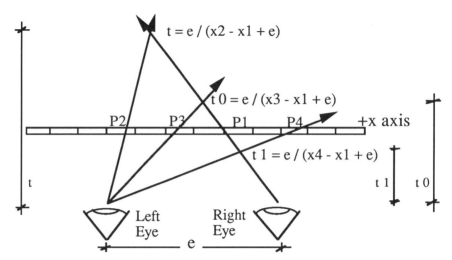

Figure 7.10
Overlapping pixel
problem.

given by $x = x_0 + t(x_1 - x_0)$, the following x values are obtained for the two rays: $x_l = -e/2+t(x_2+e/2)$ and $x_r = -e/2+t(x_1-e/2)$. Setting the two equations equal yields $t = e/(x_2 - x_1 + e)$. As shown in Figure 7.10, a point on the surface of an object will appear in P2 in the left-eye view and in P1 in the right-eye view if and only if there is an object surface at $t = e/(x_2 - x_1 + e)$. If another pixel, P3, at $(x_3, y, 0)$ satisfies the condition $x_2 < x_3 \leq x_1$, a ray from the left eye will intersect the ray from the right eye through P1 at $t_0 = e/(x_3 - x_1 + e)$. Thus, if there is an object surface at that point, it would appear in P3 in the left-eye view and P1 in the right-eye view. This causes what Badt calls an overlapped pixel problem, where more than one object is projected to the same right-eye pixel.

The reprojection of a right-eye-view pixel from the left-eye-view pixel is dependent on the z value of the image in the left-eye pixel, as can be seen from the equation for x_{sr}. Quite often the z values of two adjacent pixels in the left-eye view are such that the second left-eye pixel is reprojected more than one pixel away from the first. Thus, it may happen that sometime earlier or later in the scan line other pixels could have been or will be reprojected into this gap. As shown in Figure 7.11, the shaded pixel is a "bad" pixel since it may not have the correct pixel information for the right-eye image.

To eliminate the overlapped pixel and bad pixel problems, Adelson and Hodges suggest the use of a data structure that would hold information about the points reprojected from the left image to the right. This data structure would have an entry for every pixel in a scan line, each containing a Boolean flag, which is set to true if there is reprojection to that pixel, and a field that holds the color value for the pixel. By processing the pixels from left to right and overwriting the scan line record every time a reprojection occurs, the correct object is always reprojected to the pixels in

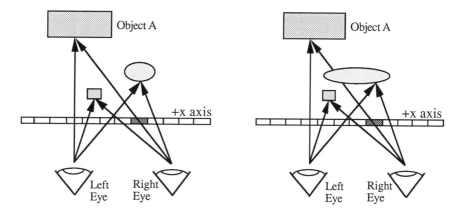

Figure 7.11
Bad pixel problem.

the right-eye view. The bad pixel problem may be solved by setting the flags of all the intervening pixels as false for the right-eye view when a jump of more than one pixel occurs. Thus, all bad pixels are forced into becoming missed pixels. They are pixels corresponding to the right-eye view that do not exist in the left-eye view and need to be ray-traced for the right eye.

7.7 Converting Algorithms to Stereo

Algorithms that are efficient for single-frame rendering may or may not be efficient for stereo rendering, or extending the algorithm to handle stereo may not be straightforward. Following is an overview of the stereo adaptations and improvements to various graphics algorithms that are presented in more detail in [ADEL91]. Following that is a detailed description of the difficulties of adapting color quantization for stereo.

Stereo Adaptation of Graphics Algorithms

Until recently, the methods for creating stereoscopic computer-generated images involved rendering each eye view separately, which on a single-processor system requires twice the work of rendering a single image. However, it is often the case that much of the work done independently for each eye view is redundant. Adelson and colleagues [ADEL91] have described ways to recognize and exploit this redundancy, thereby increasing efficiency. All of the modifications described here are based on the model using two off-axis centers of projection for computing stereo pairs described in Section 5.5. Most of the modifications are made possible by noticing from Equations 5.14 and 5.15 that the projected y value for the left eye (y_{lw}) is equal to the projected y value for the right eye (y_{rw}). We will look at scan line algorithms, clipping algorithms, and hidden-surface elimination algorithms modified for stereoscopic views.

Scan Line Algorithms. A standard scan line polygon fill algorithm makes use of an active edge list that is sorted and traversed by y-coordinate. Since the y-coordinates of the polygons in the two separate eye views do not differ, only one active edge list need be kept for both eye views. This list must keep separate data for the left-eye x-coordinates and the right-eye x-coordinates, and the polygon must be filled simultaneously in both eye views. All the work to update the active edge list can be shared, but the rest of the work must be done separately. For polygons with few sides, this means that the improvement would be negligible, but for polygons with many sides, significant work could be saved.

There is also a scan line–based hidden-surface elimination algorithm. A variation of the scan line fill algorithm, it renders pixels only if they are on front faces. This algorithm can also take advantage of a shared active edge list, but it does not achieve any other savings by simultaneously generating the two eye views.

Either the scan line fill or the hidden-surface modification can also incorporate Gouraud shading by interpolating pixel colors along a given scan line. Since the normal vector to a polygon is independent of the observer, a vertex will have the same color regardless of the viewer's position. Each vertex has the same color in the left- and right-eye views, and since the height of the polygon doesn't change between eye views, the interpolation of color for the endpoints on a given scan line can be shared between the two eye views. However, one eye view of the polygon could have more area than the other. Hence, the interpolation of color along the scan line must be done separately for each eye view.

Clipping Algorithms. The Liang-Barsky line- and polygon-clipping algorithms were explored in stereo, and the only optimization takes advantage of the fact that the y values do not change from one eye view to the other. Therefore, the y-coordinate of a line or polygon edge will be the same in both eye views, so the y parametric equation for the line need only be calculated once. Thus, clipping the line against the top or bottom borders of the screen need only be done once for the pair. Unfortunately, clipping against the left and right sides must be done separately for each eye view. The stereo Liang-Barsky line-clipping algorithm saves a considerable amount of work, since the top and bottom comparisons are a significant part of the calculations. The savings for the polygon-clipping algorithm, however, are negligible, since the computation of intersection segments, which comes after making comparisons, must be done separately for each eye view, and this calculation dominates the computations at the front end.

Hidden-Surface Elimination Algorithms. In addition to the scan line hidden-surface algorithm discussed earlier, the backface removal algorithm has been modified for stereo. The standard backface removal algorithm would compare a polygon with plane equation $Ax + By + Cz + D = 0$ with an eyepoint (X_v, Y_v, Z_v) before any perspective projection. If $AX_v + BY_v + CZ_v + D < 0$ for that eyepoint, then the polygon is a backface. If e is defined as the eyepoint separation along the x-axis, we can consider an implementation in which we have a single eyepoint and two sets of data that differ only by e in the x direction, as described in Section 5.5. Thus, for any given polygon, if the normal points generally in the positive x direction (i.e., if $A > 0$), then only one of the following three possibilities can occur. First, if the left-eye view is a backface (i.e., if $A(X_v + e) + BY_v + CZ_v + D < 0$), the polygon is a backface for both eyes, and there is no need to check the right-eye view. Second, if the left is not a backface, and $AX_v + BY_v + CZ_v + D < 0$, then the polygon is a backface for the right eye only. Otherwise, the polygon is not a backface for either position.

Similarly, if the normal points generally in the negative x direction (i.e., $A \leq 0$), then only one of the following three possibilities can occur. First, if the left-eye view is not a backface (i.e., $A(X_v + e) + BY_v + CZ_v + D > 0$), then the polygon is not a backface for either eye, and there is no need to check the right-eye view. Second, if $AX_v + BY_v + CZ_v + D > 0$, the polygon is a backface for the left eye only. Otherwise, the polygon is a backface for both positions. This algorithm saves about 33 percent of the work of doing two separate backface removal operations. In addition, if backface removal is being done as a precursor to some other hidden-surface elimination (i.e., a z-buffer), this algorithm can be modified to remove only polygons that are backfaces for both eyes, and this can be done with virtually no additional overhead for the second eye. Of course, some backfaces will not be detected, but the secondary hidden-surface algorithm will handle those few cases.

Stereo Adaptation of Color Quantization

Quantization can introduce extraneous contours in the image since there may not be enough colors to produce "smooth" shading. It can also introduce discontinuities and lack of definition in the image. We show that color quantization when applied to stereo pairs can produce noise and loss of depth in certain cases.

Practical Use of Color Quantization. Rendering is often computed using true color or 24 bits (3 bytes) per pixel. In a medium-resolution image $(1, 024 \times 1, 024)$ this equates to 3 megabytes of memory. Quantization is used to display images on graphics display devices when the frame buffer does not have a sufficient number of bit planes to represent all colors in the image, or it can be used to reduce or compress the amount of image data that must be transmitted over a network.

A color quantization algorithm selects a set of colors based on those occurring in the original image and renders the image using these colors appropriately. This set must be such that it best represents the color information in the original image using some metric or measuring technique; that is, the "difference" between the original and the quantized image must be minimized according to some algorithm for measuring differences. This set is then loaded into the color lookup table (CLUT) for the frame buffer, and the index for the CLUT entry is then used instead of the actual 24-bit pixel value.

General Color Quantization. Quantization involves three basic steps: processing the image for color information, such as counting the number of occurrences of each color (histogramming); selecting the subset of representative colors from the gamut and building the CLUT; and then rendering the image using this representative subset of colors. Heckbert's paper [HECK82] is an excellent survey of several techniques that have been proposed. A quantizer would consist of:

1. The subset of N representative colors $L = \{l_i\}, i = 1, 2, \ldots, N$. These are the colors that make up the CLUT and hence appear in the quantized image.

2. A partition of the input color space into regions called quantization cells, $R = \{r_i\}, i = 1, 2, \ldots, N$.

3. A mapping p from the input color x to representative (CLUT) indices, $p(x) = i$, if $x \in r_i$. The quantized image is made of such indices.

4. The quantization function, which maps input colors into the final colors, $q(x) = l_{p(x)}$.

Octree Color Quantization. The octree quantization algorithm [GERV88] stores the color information about the image as an octree. An octree is a tree with each node having a maximum of eight children. Each distinct color is initially entered as a leaf node. The leaf nodes are the colors that are selected for the CLUT. If the CLUT is to have N colors, then the first N different colors become leaf nodes and hence are accepted into the CLUT without any change. When the algorithm encounters the $(N + 1)$th color, it becomes the $(N+1)$th leaf node, and some leaf nodes that are close to each other are merged into a new leaf node. This is done by picking a parent node with only leaf nodes as children, replacing this parent with the average of the colors of its children, and making it the new leaf node. The representative color subset *may not contain any of the original colors* if the image is coarsely quantized.

MaxMin Color Quantization. The MaxMin [HOUL86] quantization algorithm is suited to dithering the resulting image. Here a subset of colors is chosen so that the distance $|x - y|$ between every pair (x, y) in the subset is the maximum possible, among the colors occurring in the image. The distance is measured as the sum of the squares of the RGB or HSV coordinates of the colors. If N colors are to be selected, and we have selected m colors in the subset $L = \{l_i\}$, then the $(m + 1)$th color is given by

$$\min_{i=1,m} |l_{m+1} - l_i|^2 \geq \min_{i=1,m} |x - l_i|^2 \quad \text{for every } x \text{ in } X - L$$

where X is the set of colors in the image (gamut).

Thus, the subset of colors selected in this way is well spaced in the entire gamut, and it is loaded into the CLUT. The algorithm then uses *pseudorandom dithering* to generate the appearance of colors that were in the original image but absent from the CLUT. The CLUT colors can be those that occur in the original image, but the resolution of the colors is usually reduced prior to their processing.

Color Quantization of Stereo Pairs. Quantization of stereo pairs introduces several new issues that were explored by Hebbar and McAllister [HEBB91]. One of these issues is whether to treat each of the two views as independent, and hence quantize them separately, or to treat them as one entity and quantize them together. The first approach would lead to the generation of two separate color lookup tables (CLUTs), and the second would use a single CLUT for the pair.

If each view uses a separate CLUT, then the quantization algorithm is executed twice, once for each image. As a result, it may happen that the quantized colors assigned to corresponding features in the two views may not be the same. Depending on the nature of the objects in the image, the viewer's position, and the viewing distance, it is possible to have different color histograms for the two views. Since the CLUT entries depend on the histogram, the colors in the CLUTs can be different, and corresponding features in the two views that should be assigned the same color can get different colors. This can lead to fatigue from viewing the stereo pair. Rather than combining the colors in an additive fashion, the eye perceives a shift from the color in one view to the color in the other. This phenomenon, called binocular rivalry, is described in [LEVE68] and in Chapter 4.

As an example, imagine a cube colored red on one side and green on an adjacent side. Let the image background be blue. If the viewer is so located that the left-eye view has only the red side visible, while the right-eye view has some of the green side visible too, then the number of colors in the left image is two (red and blue) and that in the right image is three (red, green, and blue). If we quantize the image to two colors in each view, the red side in the right view will have an averaged shade of either red and blue or red and green. The red side in the left view will not change color, so the red side will produce different colors in the two scenes. Fusing such a stereo pair can be difficult. Another drawback of this approach is that it effectively uses only half the maximum allowable colors in the display hardware for each view. In the previously stated example, each view had two colors, and the CLUT had a total of four colors.

If a single CLUT is used, a single set of representative colors obtained after the processing of both views is used for rendering the quantized views. Here the representative set is selected from a color histogram that has the color information of both views. In general, stereo pairs do not differ widely in their color content. Thus, there will be many colors that are common to both views. This means that each view gets more CLUT entries than would be possible with the previous approach.

For the example of the red and green cube, we have only two colors to render the image, and the problem of differing colors for a corresponding image will not occur, as both views use the same shade of red. It must also be noted here that in the previous approach, the number of entries needed in the CLUT was four, whereas in this case it is two.

The smooth shading and specular highlights on a three-dimensional object in an image act as depth cues. Such an object would be composed of many colors. Quantization of these scenes leads to a reduction in the number of colors present and hence affects the depth cues. The quantization is manifested as bands or contours in the image. In extreme cases, the object loses most of its gradual shading and shows strongly demarcated regions. Such objects tend to flatten out and appear as two-dimensional objects in the 3D space seen by the viewer. In the octree method, this effect is more pronounced for small objects whose colors make up a small percentage of the total pixels in the image. In the MaxMin method, the shading is maintained adequately until the image is heavily quantized, and there is no bias against small pixel populations. The MaxMin method does not exhibit contouring until the image is quantized very coarsely, due to its use of pseudorandom dithering. The octree method is prone to contouring.

In certain situations, quantization can lead to loss of the boundary definition and introduction of discontinuities. Let A and B be two objects, one of which partially obscures the other. Let their colors be C_A and C_B, respectively. If the distance between C_A and C_B is small, then, due to quantization, both of these colors can fall into the same quantization cell and hence get mapped to a single color. Thus, the boundary between the objects is lost. This can also happen when the image background color is close to the object color; in this case, parts of the object will merge into the background. This can create nonsolid objects from solid ones. For example, visualize a multifaceted cylinder with one of the visible facets colored green. Let the background be a shade of green that is close to the facet's color. Quantizing this image will color the facet the same as the background. When viewed in stereo, this will appear to be just two planes with some distance between them, with the background taking the place of the green facet. Colorplate 7.7 shows a stereo pair before quantization, and Colorplate 7.8 shows the same pair after quantization.

Color Quantization of Animation. In animation sequences, quantization of individual stereo pairs may cause a drift in the color of the objects. Sistare and Friedell [SIST89] have used all the possible colors that can be generated in an animation sequence to build a static CLUT to be used for the entire sequence. However, this can lead to increased quantization error in a single frame if methods similar to octree quantization are used, as each frame may not have all the colors that were used to build the CLUT. For example, if the sequence had 712 distinct colors in all, and the images were quantized to 256 colors, then an image in the quantized sequence will have more averaged colors in it than it would have if the images were individually quantized to 256 colors.

It may be possible to quantize over several frame sequences in parallel and then combine the resulting CLUTs. Methods are needed to allow weighting of specific frames to minimize differences between colors of frames at the boundaries of frame sequences, so that jump discontinuities in color do not occur.

7.8 Conclusions

We have examined several different areas in stereo system implementation, including input devices, interface concepts, and algorithm efficiencies. The improvement in technology and availability of stereo has galvanized research in the area. Issues in stereo depth perception continue to be an important topic of research. Applications such as stereo animation have motivated the development of stereo drawing devices and inspired research in such areas as data compression and algorithms for stereo motion blur. The field of stereo computer graphics will continue to be a rich area for researchers.

8

Moving-Slit Methods

Homer B. Tilton

8.1 Introduction

Moving-slit methods are relatively new on the 3D display scene. Sculpture probably represents the oldest 3D technology. Stereoscopic displays came along circa 1838, with Sir Charles Wheatstone's invention of the stereoscope. In 1932 Vladimir Zworykin filed U.S. Patent 2,107,464 for a stereo television system using a parallax barrier CRT display. In the 1940s methods of monocular 3D for CRTs were proposed by many workers, culminating in today's realistic monoscopic computer 3D. Then, in rapid succession, moving-screen and -mirror displays appeared, holography became practical with the invention of the laser, and Robert Collender announced the first moving-slit display.

Multiple-slit parallax barrier technology promised realistic 3D displays and, for the most part, they have shown that they can deliver on that promise, except possibly in the area of real-time displays. "Real time" refers to the ability of images to change smoothly or abruptly and to present motion with no visually discernible delay between the application of raw data and the presentation of the displayed image. The multiple-slit design of parallax barrier displays can be dramatically simplified in a real-time display. A real-time display must contain active, dynamic (not necessarily mechanical) systems so that the images can be continually updated, and this suggests the use of a single slit that is moving so that its effect is distributed or "time-shared" over the entire screen area. This reduced complexity can result in a dramatic decrease in cost without compromising the quality of the images.

Two embodiments of moving-slit technology will be discussed: Collender's stereoptiplexer and the parallactiscope. The focus of this chapter is on the parallactiscope. Focusing on one instrument allows the discussion of real-time displays to be much more specific about design details than would otherwise be possible. The chapter also discusses how the parallactiscope arose out of lessons learned from holography, the present state of parallactiscope development, immediate uses for the device, computer displays, television applications, and other related work.

8.2 The Stereoptiplexer

Collender [COLL67] used a single moving slit in conjunction with images projected from a moving film strip in his stereoptiplexer. Thus, his imaging medium was photographic. In the stereoptiplexer, a drum rotates about a central axis and carries a single vertical slit parallel to the axis. From below, a continuous-strip movie projector projects an image onto an internal horizontal screen located at the base of the drum. The screen image is viewed through the slit via a 45° mirror rotating inside the drum so that it continually faces the slit. By synchronizing the prerecorded film image motion to the drum rotation, the proper views are presented at the proper times so that the observer sees an opaque (solid) 3D image incorporating stereo, movement parallax, and other depth cues (see Figures 8.1a and 8.1b).

Stereoptiplexer images contain autostereoscopic and "lookaround" (motion parallax) properties just as parallax barrier displays do. The parallactic nature of the images means that multiple observers can view images simultaneously and see differing views in stereo depth.

8.3 The Parallactiscope

The parallactiscope is an oscilloscope-like moving-slit instrument that generates images having horizontal parallax on a CRT in real time. As with the stereoptiplexer, this means that images have autostereoscopic and lookaround properties. With present parallactiscopes, an observer can move up to 45° either way to see partially around the images. The parallax inherent in parallactiscope displays automatically generates the horizontal component of linear perspective. The vertical component is added by conventional means, as will be described. Present parallactiscopes have one mechanical moving component—the slit itself—but the way is clear to replacing that with a nonmechanical device. Since the parallactiscope is

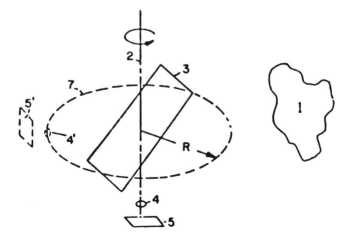

Figure 8.1a
Illustrations from
U.S. Patent
3,178,729.

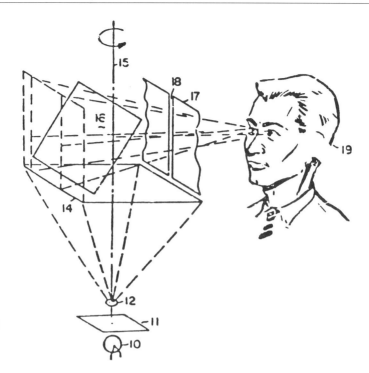

Figure 8.1b
Illustrations from
U.S. Patent
3,178,729.

essentially a new kind of oscilloscope, its images are generated the way oscilloscope images are generated, namely, by injecting analog waveforms into its three deflection inputs. The rules of parametric analytic geometry apply just as for an ordinary (2D) oscilloscope; thus, to generate a helix on the parallactiscope screen, you inject a pair of quadrature sinusoids and a sawtooth wave.

The Parallactiscope Principle

Figure 8.2 illustrates the parallactiscope principle. Different narrow vertical strips of a CRT screen are keyed to different viewing directions by use of a slit spaced in front of the screen. By moving the slit rapidly from side to side over the full width of the screen, the slit can be made to effectively disappear while its parallactic action is distributed over the entire width of the screen. In practice, a slit overscan is used so that the observers can see the entire screen even when they are not directly on axis. Such an arrangement provides a parallax barrier, as described in Chapter 3, in which the number of slits is reduced to one. This arrangement requires a dynamic imaging medium, such as a CRT screen, because the image must be continually changing.

The resulting images contain horizontal parallax, as mentioned earlier, but they do not contain vertical parallax or the accommodation (focus) cue. Because there is horizontal parallax but no vertical parallax, if an observer's

View from Above

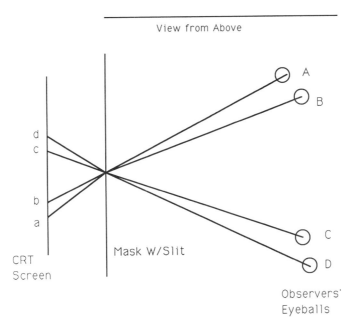

Figure 8.2
Keying different zones of a CRT screen to different angles of viewing.

head is tilted to one side, the amount of stereopsis will be reduced. Indeed, if the viewer's head is tilted left or right so far that the line joining the two pupils is vertical, there can be no stereoscopic sensation whatsoever; however, side-to-side movement parallax will still operate. The term *holoform* has been coined to describe visual properties shared by rainbow and other kinds of holograms, parallax-panoramagrams, parallax barrier displays, the stereoptiplexer, and the parallactiscope. The most important of these properties are:

> Linear perspective
> Autostereopsis (head straight or nearly so)
> The lookaround property
> Solid and wireframe images

The last property has yet to be demonstrated with the parallactiscope.

The parallactiscope principle was introduced in a 1971 publication [TILT71]. A working model was first demonstrated publicly in 1977 [TILT77]. Since that time, public demonstrations have been given of increasingly complex hardware ([TILT82], [TILT85], [TILT88]). A hardcover book describing how to build and use the parallactiscope appeared in 1987 [TILT87].

The Parallax Transformation

The parallactiscope principle can be applied to a variety of hardware designs, but perhaps the simplest application is an oscillographic direct-writing approach, as is used in the parallactiscope. Just as an ordinary (2D) oscilloscope generates flat images (two-dimensional spaceforms) defined by a pair of deflection signals (waveforms), a 3D oscilloscope can generate three-dimensional spaceforms defined by a trio of deflection signals. Figures 8.3 and 8.4 illustrate the imaging geometry involved in such a 3D oscilloscope, or parallactiscope. The view shown in Figure 8.3 leads to Equation (8.1), the horizontal parallax transformation. The view shown in Figure 8.4 leads to Equation (8.2), the vertical perspective transformation. The vertical perspective transformation applies to a wide variety of 3D imaging systems. The horizontal parallax transformation is peculiar to the parallactiscope principle.

$$h = \frac{x - s}{1 - z/a} + s \tag{8.1}$$

$$v = \frac{y}{1 - z/A} \tag{8.2}$$

Equation (8.1) reveals how one must process the horizontal (x) and depth (z) signals to produce the horizontal deflection signal (h) required to produce holoform images. The instant-to-instant slit position (s) is a parametric input. The slit constant (a) reflects the distance of the slit from the CRT screen. In Equation (8.2), a nominal observer distance (A) is used to determine perspective requirements for the vertical deflection (v) signal from the vertical (y) and depth (z) signals. In the parallactiscope, the horizon-

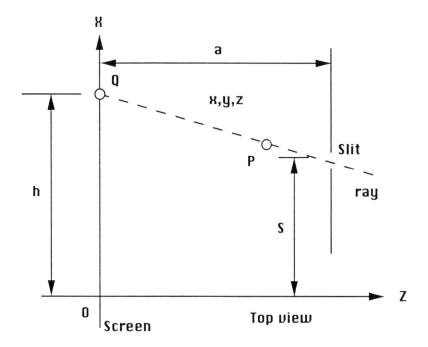

Figure 8.3
Geometry for the horizontal parallax equation.

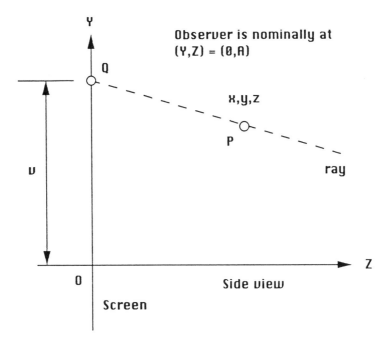

Figure 8.4
Geometry for the vertical perspective equation.

tal parallax transformation and the vertical perspective transformation are performed by an analog instrument called a parallax processor.

Parallax Processor

The block diagram of the parallax processor is shown in Figure 8.5. This diagram is simply a reflection of Equations (8.1) and (8.2). If the block diagram is implemented by high-speed analog electronic circuitry, the required transformations described by Equations (8.1) and (8.2) will be performed on the three input signals in real time. To determine the form that circuitry takes, a schematic diagram is needed. Developed from the block diagram, it is shown in Figure 8.6. The resulting hardware, contained in a box measuring $3\frac{1}{2} \times 8 \times 11$ inches, produces the signals required by an ordinary oscilloscope to produce holoform images in accordance with the parallactiscope principle.

A vertical slit must be provided, and the slit must be scanned rapidly from side to side. The slit's instantaneous position must also be continually reported to the parallax processor. The moving slit is contained in a device called the parallax scanner. The slit is driven by the electrical output of another device, called the scanner driver, and a slit position signal is also generated there. These two items are discussed next.

Parallax Scanner

The ideal moving slit would be all electronic, perhaps involving a variety of liquid-crystal technology. Lacking that, the slit should be lightweight so

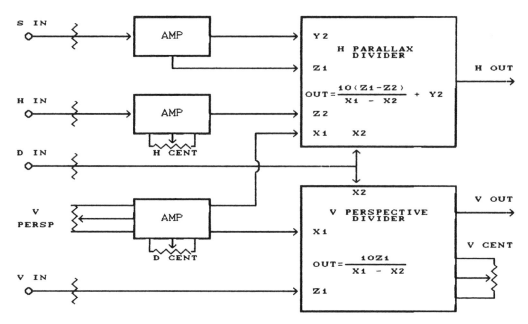

Figure 8.5 Block diagram of the parallax processor.

that it can be moved rapidly with very little energy expended. This goal can be realized by using a sliver of half-wave retarder placed between crossed linear polarizing sheets. Only that sliver of plastic needs to be scanned in this arrangement. The physical slit (sliver of plastic) produces an optical slit in an otherwise dark, opaque mask. In the parallactiscopes built to date, the slit is placed on the end of a lightweight arm and caused to oscillate back and forth in a sinusoidal fashion. An ordinary loudspeaker provides the motive force. The arm is configured as a torsion pendulum and is made to oscillate at its resonant frequency. In this way, very little energy is required to keep it oscillating once it is started. (It is self-starting.) Present models have a resonant frequency of about 12 Hz, providing about twenty-four slit scans per second; thus, there are two slit scans for each cycle of oscillation—one with the slit moving rightward and one with it moving leftward. There is no problem obtaining peak-to-peak (double-amplitude) slit scans that exceed the screen width (overscan condition). The parallax scanner sits atop the host oscilloscope with its slit spaced in front of the CRT screen by about half the width of the screen.

Scanner Driver

The oscillator circuit that provides the driving power for the loudspeaker voice coil in the parallax scanner also provides the s signal required by the parallax processor. That signal corresponds to the s parameter in Equation (8.1). Because of the simple harmonic motion of the slit, voice coil motion is naturally sinusoidal. Therefore, sinewave signals are involved. The

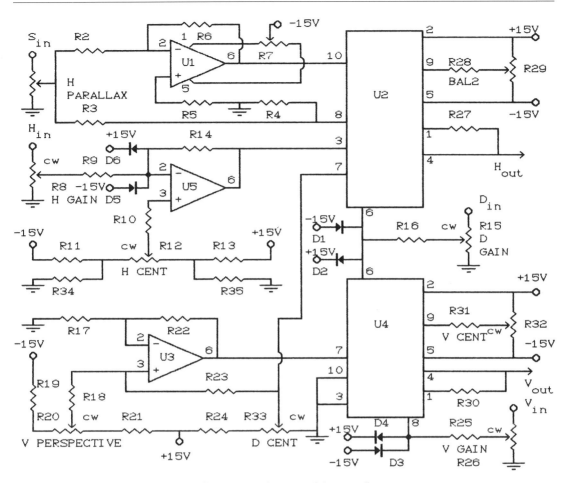

Figure 8.6 Schematic diagram of the parallax processor.

voltage component of the sinewave power to the voice coil and the sinewave signal voltage (*s* signal) to the parallax processor must be 90° out of phase to compensate for the 90° phase shift produced by voice coil inductance. Therefore, a quadrature sinewave oscillator is used in the scanner driver. In present models, this oscillator runs as an open-loop (free-running, unsynchronized) oscillator. In original models, the pendulum was part of a feedback loop so that it governed the frequency of the oscillator. It was thought that this would ensure a phaselock condition so that the images produced for each direction of slit scan would be exactly superimposed. In practice this has not worked out, so the simpler open-loop design is presently used. With either approach there is a minor problem of manually synchronizing the driving oscillator with the pendulum resonant frequency so that the two images generated rightward and leftward will be superimposed. With the advent of an all-electronic scanning slit, it is anticipated that this problem will resolve itself.

Construction

The parallactiscope has been called "Everyman's real-time real 3-D" [TILT89] because virtually any electronic tinkerer with access to an oscilloscope can build a parallactiscope for a parts cost of less than $500, exclusive of the oscilloscope cost. High-precision fabrication techniques are not required, largely as a result of the use of a single slit having simple harmonic motion. The oscilloscope needs to be a dual-trace, dc-coupled type having a short-persistence phosphor such as EIA P31 or WW GH. Except for removal of any protrusions on top of the case, the oscilloscope does not need modification. Although enough information is contained in this chapter to enable a determined person to build a parallactiscope, there is not enough space for the level of detail that a book devoted to the instrument can provide. For that level of detail see [TILT87].

The Elusive Real-Time Hologram

The hologram performs its visual magic through ray reconstruction, which it performs by means of wavefront interference. This not only reconstructs ray directions; it also reconstructs phase fronts. But phase front information plays no role in visual space perception, and it is not needed for realistic 3D displays. Indeed, ray reconstruction by other means might make speckle and color properties more manageable. For example, parallactiscope images do not contain speckle, and a color CRT can be used to produce color images. The required short-persistence blue and green phosphors are currently available, but a suitable short-persistence red phosphor may require further development.

The parallactiscope, like other slit-based displays, is not holographic. It does not reconstruct ray directions by using wavefront interference, but it does construct or control ray directions—at least their horizontal component—by other means. And ray reconstruction (rather than wavefront interference per se) is the mechanism underlying the visual magic of the hologram. The type of ray reconstruction used in fixed- and moving-slit displays is unlike that used in holograms in that phase fronts are not reconstructed, but it is similar in that such techniques are well suited to displaying opaque (solid) objects as well as transparent (wireframe) objects.

8.4 The Present State of the Art

Brightness with the parallactiscope is good because a very narrow slit is not required. Present parallactiscopes use slit widths from $\frac{1}{10}$ to $\frac{1}{8}$ inch (about 2 to 3 mm). The slit reduces the physical intensity of the light by the ratio of the slit width to the total scan width. But recall that neither photographic brightness nor visual brightness is linearly related to the physical intensity of light; the function is very nearly logarithmic in this range of intensities. Thus, the reduction in visual brightness is not that much. The presence of the slit causes a reduction in brightness, compared to a "naked" CRT screen, of about 5:1 in terms of photographic exposure times required for the two cases. Thus, typical exposure times for taking scope photos are about 1 second with the slit structure removed and 5 seconds with it in place and operating normally. In this range of intensities, visual (subjective) brightness and photographic brightness are closely related.

How many different and discrete images have to be generated for each complete scan of the slit? That question applies to a digital system, but not to an analog system. A digital imaging system may require as many different and discrete images as there are different and discrete slit positions. This may seem to be a very severe requirement, but with an analog system the requirement is easy to meet. In an analog system, "different and discrete" images are not presented. Instead, there is a continual, smooth transition from view to view as time flows. That is, as the spot is traced out on the CRT screen, its path is continually adjusted by the depth signal. Thus, we are really talking about only one dynamic image. To appreciate this dynamism, one need only view a parallactiscope image without the slit present, but with it still running. The image rapidly pulsates horizontally in step with the slit oscillation.

While the preceding question may not apply to an analog system, a related question does. Is there any particular range of signal frequencies that should be avoided? There is a band of frequencies, centered in a geometric sense on the slit scan frequency, that might be called the Moiré band. If the fundamental frequency of any of the three input signals is in the Moiré band, the image will break up; but unless one of the three signal frequencies is actually synchronized with the slit scan frequency, there will still be an image having integrity in the sense that a photographic time exposure of a "stationary" image would show it as a normal image. Such an image is, however, difficult to view directly because it is built up over many scans in a piecewise fashion. The Moiré band runs from about 1 to 1,000 Hz. Thus, one needs to avoid signal frequencies in that band for direct viewing.

An outboard analog rotator can be attached to the input terminals of the parallactiscope to permit the viewer to manipulate the displayed image in 3-space. Such rotators are an integral part of a precursor of the parallactiscope, the Optical Electronics "scenoscope," which won an I·R 100 award in 1965. The scenoscope is a special-purpose oscilloscope that displays real-time scenographic projections, or "monocular 3D," of a kind referred to simply as "3D" in today's computer jargon. Analog rotators are old technology and have not yet been demonstrated with the parallactiscope.

Immediate Uses

Even in its present form, the parallactiscope has some immediate practical uses. "Multiscopic" photos of displays are shown in Figures 8.7 through 8.11. Multiscopic photos are like stereoscopic ones except there are more

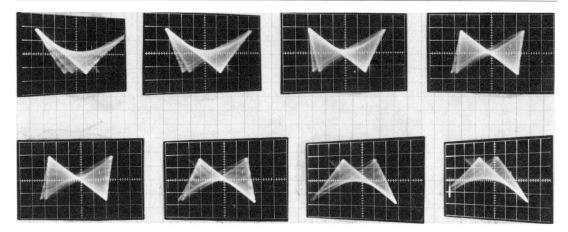

Figure 8.7 Multiscopic views of a hyperbolic paraboloid.

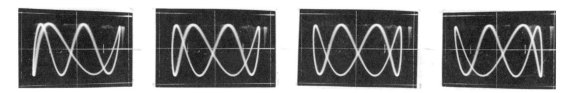

Figure 8.8 Multiscopic views of a 3D Lissajous figure.

than two, with each adjacent pair stereoscopically related. Viewing such photos conveys some of the feeling for the live displays, because moving to an adjacent pair of photos allows you to see partly around the displayed object.

As an instrument for use in the electronics laboratory, the parallactiscope can display three signals as a three-dimensional surface, curve, or "waveform" for analysis. For example, the characteristic surface of a four-quadrant analog multiplier is a hyperbolic paraboloid (saddle-shaped surface). This can be generated by applying two sinusoids of widely different frequencies to the multiplier's two inputs. Such patterns can be used to give a quick indication of the electrical integrity of analog multipliers. The signal combination of two quadrature sinusoids with a sawtooth generates a helix (spiral) or cylinder, depending on frequency ratios.

The spiral, a twisted space curve, is one of the easiest figures to generate on the parallactiscope. Indeed, a spiral may have been the first space curve ever generated by a CRT in a "truly" 3D way, as reported in a 1921 paper [HULL21], which stated, "The bundle of 'rays' emanating from the filament cathode and made visible by the ionized air (inside the CRT) can be formed into beautiful spiral figures by the action of the focussing coils" (p. 140).

As an educational instrument, the parallactiscope can—at low cost—display a wide range of user-synthesized real-time holoform images, includ-

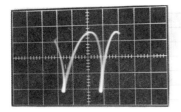 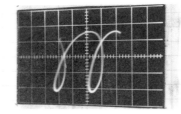 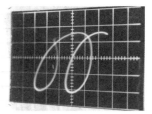

Figure 8.9 Multiscopic views of a helix.

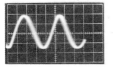 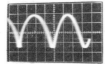 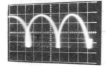

Figure 8.10 Multiscopic views of a helix.

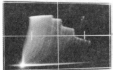  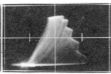 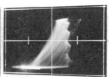 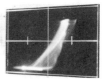

Figure 8.11 Multiscopic views of an abstract figure.

ing such classic three-dimensional figures as spheres, cones, ellipsoids, hyperboloids, elliptic paraboloids, and hyperbolic paraboloids. It is a dramatic visual experience to view a surface that slowly metamorphizes, under manual control, from a cone of two nappes to a sphere and back again. This can be accomplished by a simple analog circuit that the student can assemble. Details on how to display those figures and others are given elsewhere [TILT87]. These methods draw directly from the equations of solid analytic geometry.

Finally, there is the intriguing idea that by manipulating 3D projections of 4D figures in a real-time holoform display, one might—with practice—eventually develop a visual sense of the four-dimensional form of those figures. Noll reported on experiments of this type using computer-generated displays in 1968. For the latest on this subject, and for some interesting computer-generated line drawings of tesseracts, see Armstrong [ARMS90]. Such experiments are easily performed using the parallactiscope. It is important to realize that it is possible, using techniques such as those reported here, to actually manipulate (rotate) virtual 4D figures in 4-space even before their 3D projections are generated! Experiments of this nature using conventional 3D on a digital computer have shown negative results, perhaps mainly because of the difficulty of close real-time user interaction with the displayed figures in such an arrangement. Just as a slowly changing 2D

Lissajous figure suggests depth, displaying a slowly changing 3D Lissajous figure on the parallactiscope might suggest a fourth metric extension.

8.5 Television and Other Work

Ian Sexton [SEXT89] is currently engaged in an attempt to apply the parallactiscope principle to television images. In the 3D TV arena, one might envision a display much like the parallactiscope but using a raster instead of a vector presentation. This display would use a narrow vertical raster that moves so as to always be opposite the slit position. To produce the stereo cue, this narrow "instantaneous" raster must be at least wide enough to encompass the positions of both eyes of a single observer viewing through the slit. To produce the lookaround property as well, the raster must be even wider. How wide depends on how far one wants to look around.

Image generation (sensing) becomes another problem. One approach is to use a camera that generates a depth video signal in addition to the one (or three, in a color system) normal-intensity video signal. The reconstituted image on the CRT screen would use that depth signal to generate a very wide-angle view. Two cameras using a triangulation method might do the job.

A recent patent [HATT90] describes a closed-circuit television format using systems of moving pinholes or slits and moving lenses. A very wide-aperture lens is used on the camera, so that only a single stationary camera is required. Thus, the range of horizontal motion parallax available to the viewers is a function of the camera's width and distance from the scene.

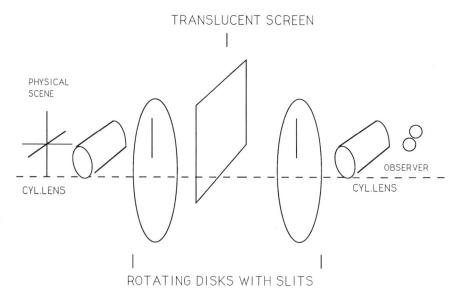

Figure 8.12
Mockup of a parallactiscope TV demonstrator using two 12-inch phonograph records to which slits have been added.

TRANSLUCENT SCREEN

PHYSICAL SCENE

CYL.LENS

OBSERVER

CYL.LENS

ROTATING DISKS WITH SLITS

However, the experimental apparatus used to demonstrate the principle apparently consisted of a camera that actually moved from side to side.

An even simpler device to illustrate the principle can be made from two 12-inch phonograph records carrying single aligned slits and rotating rapidly on the same horizontal shaft, with a ground-glass screen halfway between them. Looking through one record as it rapidly rotates is like looking directly at an object behind the other record, except that the image you actually see is the dynamic two-dimensional image formed on the screen (see Figure 8.12). This mockup demonstrates how a three-dimensional image can be coded onto a two-dimensional screen and reconstituted without requiring observers to wear special apparatus or take up special positions.

Finally, in other related work, Lowell Noble [NOBL87] has added his virtual-imaging lens system to the front end of a parallactiscope. Joel Kollin [KOLL88] has proposed the use of rocking slats or louvers instead of a moving slit to impart parallax to real-time images. Finally, Adrian Travis [TRAVn.d.] is presently working with slits composed of liquid-crystal substances.

9

The Parallax Illumination
Autostereoscopic Method

Jessie Eichenlaub

9.1 Introduction

Dimension Technologies, Incorporated (DTI), of Rochester, New York, is currently manufacturing autostereoscopic displays employing a unique, proprietary optical technique. The basic technique, which is called parallax illumination, is very simple, is easy to implement, and produces very vivid stereoscopic images.

Parallax Illumination was invented in the mid-1980s by the author and has since been covered by a series of issued and pending patents. The company was incorporated in 1986 to develop the technology into useful products. The first product, a black-and-white autostereoscopic display, was introduced in 1989, and a color version was introduced in 1992. Figure 9.1 is an optical diagram of the simplest version of the system as viewed from the top, and Figure 9.2 is a perspective close-up view.

The DTI system employs a transmissive image-forming display, such as an LCD, situated in front of and spaced apart from a special illumination plate [BUSQ90]. The illumination plate produces a large number of thin, bright vertical illuminating lines with dark spaces in between. There is one line for every two columns of pixels. The lines are spaced such that an observer sitting at the average viewing distance from the display sees all of the light lines through the odd columns of pixels with the left eye and the same set of lines through the even columns with the right eye [EICH90]. Figure 9.3 shows magnified views of what the display looks like to each eye.

Since the display is transmissive, the information on any pixel can be seen only when illumination is seen behind the pixel. Thus, the observer's left eye sees only the information on the odd columns of pixels, and his or her right eye sees only what is on the even columns. To produce a stereoscopic image, the left-eye view of a stereo pair must be displayed on the odd columns and the right-eye view on the even columns.

Figure 9.4 further illustrates the geometry of the system. A left-eye view is seen anywhere within the quadrilateral-shaped areas marked L, and a right-eye view is seen anywhere within the areas marked R. As can be seen,

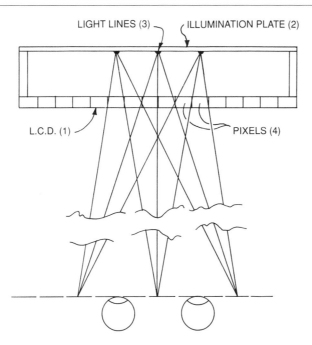

LIGHT LINES (3) ILLUMINATION PLATE (2)

L.C.D. (1) PIXELS (4)

Figure 9.1
Optical diagram
of DTI
autostereoscopic
system (top
view).

there are several left- and right-eye viewing zones in front of the display. These zones are all widest at a certain plane that is parallel to the display surface and situated at the most comfortable viewing distance from it. Thus, multiple observers can position themselves next to each other to perceive stereoscopic images.

The positioning of the lines can be calculated with simple geometry in a two-step process. First, the spacing between the lines and the pixels should be calculated.

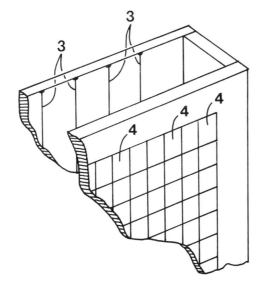

Figure 9.2
Perspective
close-up view
of DTI
autostereoscopic
system.

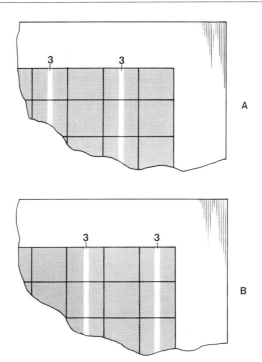

Figure 9.3
Left-eye (a) and right-eye (b) views of the display.

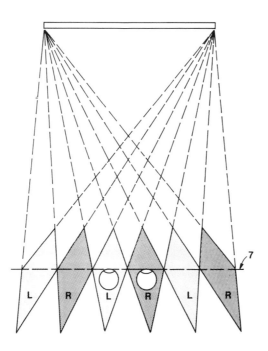

Figure 9.4
Viewing zones produced by the autostereoscopic display.

The ideal width of a viewing zone at the widest point is 63 mm, the average distance between a person's eyes. This width allows the average person the greatest freedom of head movement from side to side. As can be seen in Figure 9.5, a number of left-eye and right-eye zones are present, with thin dead zones [FOLE90] between them. In these zones, the light lines are seen behind the spaces in between the pixels. Increasing the zone width also increases the width of the dead zone and thus narrows the distance in which the person can move left or right before one of the eyes crosses into a dead zone.

Likewise, narrowing the zone width also narrows the dead zone width but moves the dead zones at the outer edges inward, which again reduces the distance that a person can move before one of his or her eyes crosses the outer-edge dead zone. Thus, for any given observer, it is best to have zones that are equal in width to the center-to-center distance between the pupils. Again, this distance is, on the average, 63 mm, which seems to provide a comfortable zone for most adults.

The designer must specify the ideal viewing distance for a particular display. A typical viewing distance is about 30 inches, but depending on the screen size and application, this could vary. In any case, the spacing between the lines and the pixels can be calculated through simple geometry from Figure 9.4, using the following parameters:

S = illuminating line pitch

P = pixel width

D = optical distance between the illuminating lines and the pixels of the transmissive display

V = viewing distance

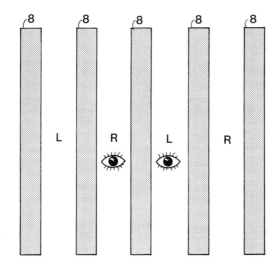

Figure 9.5
Viewing zones of the autostereoscopic display.

Because of similar triangles,

$$\frac{P}{D} = \frac{63}{V + D}$$

or

$$D = \frac{V \times P}{63 - P}$$

Given this distance, the pitch between the illuminating lines can be calculated. It can be proven by the similar triangles method that

$$\frac{S}{V + D} = \frac{2P}{V}$$

or

$$S = \frac{2P(V + D)}{V}$$

This formula shows that the distance between adjacent illuminating lines is equal across the illumination panel and is slightly greater than twice the pitch of the display pixels.

There are many ways to produce a series of thin, bright lines on a black background. In DTI's products, a bright linear light source is placed along one side of the display. An inexpensive aperture fluorescent lamp of the type employed in some copiers is currently used. Light from the aperture is collimated by a cylindrical lens and made to fall across a special reflector plate. The plate possesses a large number of thin V-shaped ridges on its rear surface, parallel to the short dimension of the display. The ridges are spaced at the light line separation distance specified in the preceding formulas. The ridge height is chosen such that for a perfectly collimated beam of light, any light that just passes one ridge enters the next, and so on across the plate, so that little light is wasted. The sides of the ridges are angled so that light entering one side is reflected by the far side, due to total internal reflection, in a direction toward the reflector plate surface and perpendicular to it. The light then strikes a diffusing layer and is scattered. An observer in front of the display sees a large number of bright, thin, precisely spaced vertical light-emitting lines on the diffusing layer, with dark spaces in between.

One versatility aspect of this illumination arrangement is that it can be used for full-resolution 2D as well as 3D viewing. If the reflector plate illumination source is turned off and a conventional evenly distributed light source, such as an EL panel, is turned on behind the reflector plate, the observer will see all the pixels with both eyes and can use the system at full resolution for conventional 2D applications. Looking at nonstereoscopic images with the 3D illumination on can be annoying, since each eye sees different pixel columns of the same image. In the case of text, for example, each eye sees different parts of each letter. A 2D viewing feature is therefore very important when one display must be used for several applications, and not just stereo viewing.

The system is not limited to use with an LCD possessing pixels in straight rows and columns. It can be adapted to color LCDs with pixels arranged in a triad configuration. This is done with a special opaque checkerboard mask situated on or near the LC layer, as illustrated in Figure 9.6. In this case, the light-emitting lines [CHER90a] are spaced at such a distance that there appears to be one line behind every pixel of the LCD. The opaque sections of the mask [GESC90] cover half of every pixel, and the overall mask pattern is arranged so that to the left eye the light lines are visible through the odd rows of triads and are hidden by the mask on the other rows. Likewise, to the right eye the lines are visible through the mask on the even rows of triads but are hidden by the mask on the odd rows. It is interesting to note that this sort of arrangement, where each eye sees every other row of the display instead of every other column, makes the display compatible with the software and camera systems used with many of the "winking" 3D glasses systems commonly used for home VCRs and video games, since these interlace the left- and right-eye views in adjacent horizontal rows. Masks with different configurations can be made to deal with various color stripe arrangements or pixel quad arrangements.

9.2 Head Tracking with a Parallax Illumination Display

All simple types of autostereoscopic displays produce head-position restrictions. One method of decreasing these restrictions, at least for a single viewer, is head tracking. Head tracking is an old idea that involves sensing the position of the observer's head and adjusting the optics of the autostereo-

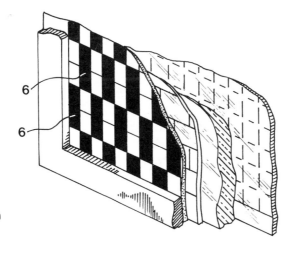

Figure 9.6
Checkerboard mask for use with pixel triads.

scopic display so that the viewing zones stay positioned at the observer's eye locations.

Dimension Technologies has experimented with a head tracker that uses two ultrasonic transducers positioned at the top corners of a display to sense the position of a single observer's head. A computer measures the time it takes for pulses emitted from the transducers to bounce off the observer's head and return, and from this it calculates the distance between the observer and each transducer, and thus the observer's head position relative to the screen. The illuminating lines can be moved sideways, by at most 1 mm in either direction, to shift the position of the viewing zones by about 300 mm to either side and thus keep the left-eye and right-eye zones centered near the observer's eyes. Movement of the lines can be accomplished electromechanically or electro-optically.

"Lookaround" Capability

A head-tracking system can also be used to provide a "lookaround" effect for a single observer. When computer-generated images are being viewed, the head tracker can return the head position to the image-generating program, which in turn can calculate estimated eye positions and continuously redraw the left and right perspective views from eyepoints congruent with the observer's eye positions at the time of the redraw. Thus, as the observer moves from side to side and even forward and backward, the scene will seem to change in perspective just as a real object would.

The lookaround capability just described is more than just an attractive viewing feature. It can alleviate a problem associated with stereoscopic displays—the fact that an image may appear distorted when viewed from anywhere except a single location in front of the screen.

Lookaround without Head Tracking

Given a very high-resolution LCD, it is possible to create hologram-like images with a lookaround feature without using a head tracker. Such a feature could be implemented using the arrangement shown in Figures 9.7 and 9.8. Figure 9.7 is a magnified view of part of the display, and Figure 9.8 is a perspective view of the whole display and its viewing zones. In Figure 9.7 a large number of pixel columns (columns 11–16) are situated in front of each light line [CHER90a], and the pixels are lined up so that the lines are seen through different columns of pixels from within each of the several viewing zones (zones 17–22 in Figure 9.8), spaced across a plane in front of the display. To create an image with lookaround, a view of a scene with perspective appropriate to viewing from each zone would be displayed only on those pixel columns visible from that zone. Within the first zone, for instance, a view of an object as seen from the left would be visible. Within the last zone, a view of the same object as seen from the right would be visible. As long as the zones are spaced so that an observer always has one

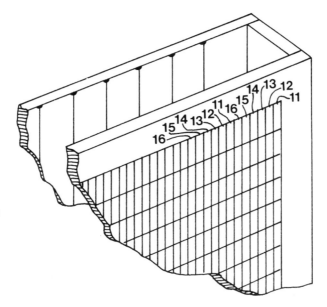

Figure 9.7
Autostereoscopic display for producing hologram-like images (top view).

eye in one zone and the other eye in another zone, the observer will see a stereoscopic image that changes perspective as he or she moves from side to side.

Of course, the disadvantage of this system is that it requires an extremely high-resolution transmissive display, since the pixel columns must be divided between several different views. If the display has M (vertical) by

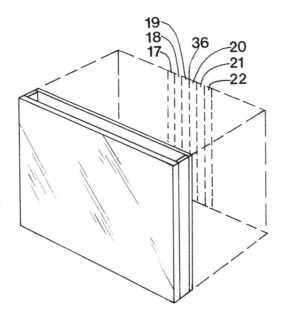

Figure 9.8
Autostereoscopic display for producing hologram-like images (perspective view).

N (horizontal) resolution and produces Z different images for Z different zones, the resolution of each image will be $M \times N/Z$. As will be seen later, there are two better ways to provide lookaround without resorting to ultrahigh-resolution displays or sacrificing resolution.

9.3 Autostereoscopy and Lookaround without Loss of Resolution

The illumination system that has been described is only the simplest version of a very versatile autostereoscopic technology. By using multiple sets of blinking light lines, different images can be made visible in different zones without any resolution sacrifice.

For example, the use of two sets of alternately blinking light lines, and time-multiplexed images on the LCD, allows both of the observer's eyes to see all of the pixels of the display (instead of half of them, as is the case with the previous configurations) and yet still see two different images. The method is illustrated in Figures 9.9 and 9.10. Figure 9.9 is a top view, and Figure 9.10 shows two magnified views as seen by the observer's right eye.

Here two sets of light lines are used, marked 24 and 25. Lines of the second set are spaced halfway between the lines of the first set. The two sets blink on and off very rapidly in succession, so that to an observer in front of the display, the lines seem to jump back and forth between the two

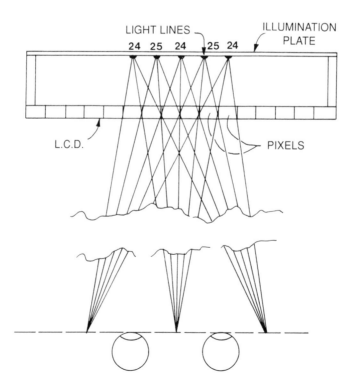

Figure 9.9
Full-resolution autostereoscopic display.

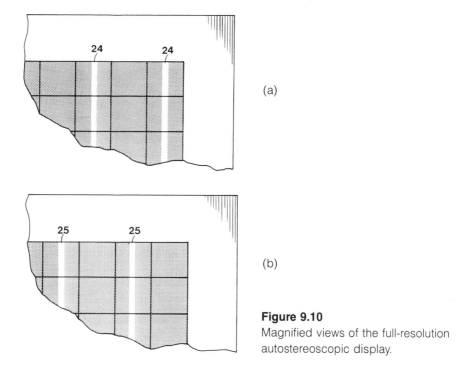

Figure 9.10
Magnified views of the full-resolution autostereoscopic display.

sets of positions. Ideally, the complete cycle of both sets flashing on and off should be accomplished in one-thirtieth of a second or less to avoid unacceptable levels of flicker.

The LCD is made to display the left-eye view of a stereoscopic pair on the odd columns when line set 24 is on, and a right-eye view on the even columns. As with the simple version of the display, an observer sees the lines of the first set through the odd columns of pixels with the left eye and the even columns with the right eye (see Figure 9.10a, showing the display as it looks to the right eye when line set 24 is on). After that light set goes out, the display is given just enough time to change the images, so that the remainder of the full-resolution left-eye view is generated on the even columns and the remainder of the right-eye view is displayed on the odd columns. When the second set flashes on, the observer now sees light lines through the even columns of pixels with the left eye, and through the odd columns with the right eye (see Figure 9.10b, showing the display as it looks to the observer's right eye when line set 25 is on). Thus, each eye sees the remainder of its respective image on the pixels that were previously invisible. During each flashing cycle, each eye sees all the pixels, but at different instants each eye sees a different set of pixels.

By using multiple sets of illuminating lines, and also increasing the frame rate of the transmissive display, one can produce more than two viewing zones, from each of which a full-resolution image with proper perspective is visible. Thus, one can create a lookaround effect without reducing the apparent resolution of each image below the resolution of the transmissive display.

The operation of a four-zone system is shown in Figures 9.11 and 9.12. Figure 9.11 is a top view, and Figure 9.12 is a close-up view of the display as it is seen from one of the viewing zones. Here four different sets of light lines are used (30–33). The members of each set are spaced apart by a little more than the width of four pixel columns, so that light from each set passes through four columns (34–37) in front of it, to four viewing zones (38–41) spaced evenly in front of the display. The sets flash on and off successively, so that at any instant only one set of lines is on. First set 30 flashes, then set 31, then 32, then 33, and then the cycle starts over with set 30. Ideally, the entire flash cycle should be completed in one-thirtieth of a second or less. When lines of set 30 are on, they are visible through pixel columns 34 from any point within zone 38, they are visible through columns 35 from any point within zone 39, and so on. While set 30 is on, pixel columns 34 are made to display part of a perspective view of a scene as it should look from the center of zone 38, columns 35 display a slightly different perspective view of the same scene as it should appear from the center of zone 39, columns 36 display part of the scene as it should appear from the center of zone 40, and columns 37 display

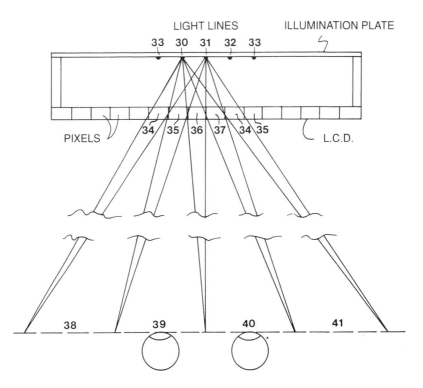

Figure 9.11
Four-zone
full-resolution
autostereoscopic
display.

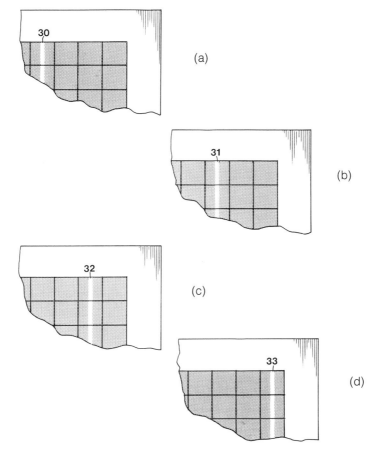

Figure 9.12
Magnified views
of the four-zone
full-resolution
autostereoscopic
display.

part of the scene as it should appear from the center of zone 41. Figure 9.12a shows how the display looks from zone 38 when lines 30 are on.

When lines 30 turn off and lines 31 turn on, these new lines are visible through columns 35 from zone 38, through columns 36 from zone 39, through columns 37 from zone 40, and through columns 34 from zone 41. Thus, between the time that line set 30 turns off and set 31 turns on, the image on the LCD changes, so that when set 31 comes on, columns 35 are displaying new parts of the perspective view appropriate to zone 38, columns 36 are displaying parts of the image appropriate to zone 39, columns 37 are displaying parts of the image appropriate to zone 40, and columns 34 are displaying parts of the image appropriate to zone 41. Figure 9.12b shows how the display looks from zone 38 when lines 31 are on.

As successive sets of lines come on, an observer's eye in zone 38 will see the lines successively through all the pixel columns 34, 35, 36, and 37, as shown in Figure 9.12. Thus, through each successive cycle, the eye sees a light line behind all the columns of pixels, and as the LCD changes, a complete full-resolution image is built up column by column. Likewise, an eye in zone 39 will see the light lines appear successively through columns 35, 36, 37, and 34; an eye in zone 40 will see the lines appear behind columns 36, 37, 34, and 35; and an eye in zone 41 will see the lines successively behind columns 37, 34, 35, and 36. Thus, a full-resolution image will be visible from each zone, and each will be a perspective view appropriate to the zone from which it is seen.

Several viewing zones must be used to avoid a "choppy" effect where an image suddenly jumps from one perspective view to another as the observer moves across the zones. Fortunately, the number of zones does not have to be excessively large; still 3D photographs using lenticular lens sheets or slit masks can achieve smooth lookaround effects with eight to sixteen zones, depending on the pitch of the slits or lenses, the viewing distance, and the apparent distance between the plane of the photograph and points on the image.

As with the simplest version of the technology, multiple observers could view stereo with a lookaround system. A large number of zones would cover a wide swath of space in front of the screen, allowing a small group of people to sit and stand within a viewing zone and see stereo without distortion. If the number of zones were smaller, the central viewing zones would not cover as much space, but they would be repeated to each side, again allowing multiple observers to see stereo.

The number of zones that can be produced, and thus the width of the viewing areas and the smoothness of lookaround, depend on the speed of the LCD or other transmissive display. The LCD must be capable of addressing its pixels and changing their state completely within a very short period of time. There are many LCD technologies that hold promise for very fast image display and change. LC materials that change their state within a few milliseconds are not uncommon, and displays made with such materials could be driven at the required rates using standard parallel driving techniques, where different segments of the LCD are driven at the same time. The fastest LCDs currently on the market are low-resolution (256×256) ferroelectric types used in optical computing research, which have frame rates of up to 300 Hz and response times of 0.1 ms. Much larger, higher-resolution fast LCDs exist in laboratories. An eight-zone display therefore is entirely within the realm of feasibility, though a practical high-resolution embodiment for typical stereo applications will require a custom LCD and bandwidths that are much higher than normal.

There are various methods of optically generating sets of light lines that flash at the required rate. The flash rates and flash durations needed are well within the limits of several lighting technologies.

9.4 Input Methods

A stereoscopic system employing any of the optical variations described so far could accept input from a wide variety of devices, including live TV cameras and computers either working alone or with input from equipment such as electronic measuring devices, radar, sonar, or medical equipment such as NMR scanners. In either case, sections of the different perspective views have to be generated on the correct columns of pixels at the correct time.

In the case of computer-generated imagery, this can be accomplished with special interface boards working in conjunction with software designed to generate the appropriate perspective views. In most cases, a software package that is designed to draw 3D perspective views on a normal 2D screen can be modified to generate the multiple views necessary for autostereoscopic viewing. Today even personal computers have the necessary power and speed to generate such views in a reasonable amount of time. In the case of stand-alone live television, electronics must be provided to accept input from two or more TV cameras, one for each viewing zone, placed side by side. The electronics must multiplex the signals so that the image from each camera is displayed on the correct pixel columns at the correct times.

9.5 DTI's Current Products

DTI is currently selling a full-color VGA monitor based on its technology. Called the DTI 1000C, this display uses a 640×480 color active-matrix LCD to provide 320×480 resolution to each eye in 3D mode and 640×480 resolution to both eyes in 2D mode. The display area is roughly 10 inches diagonal, and it can display over 100,000 colors.

The LCD contrast ratio is about 100:1, but the apparent contrast is actually higher, due to the fact that the observer sees thin bright lines with wider areas of dark space between them. The effect is similar to that achieved on most color picture tubes, where contrast is increased by surrounding the color dots with dark space.

The display is designed to operate off of any of the IBM PC or Apple Macintosh families of computers, and some workstations. It can also accept input from two multiplexed TV cameras and can show stereo videotapes off of a standard VCR. A head-tracking version of the 1000C has also been built.

DTI 1000C displays, and older black-and-white versions, are now being used in a wide variety of scientific visualization, educational, remote vehicle, remote manipulation, and experimental military applications.

9.6 Advantages of the Parallax Illumination Method over Other Autostereoscopic Technologies

The parallax illumination techniques possess several advantages over previously developed flat-panel autostereoscopic techniques, namely, the parallax barrier and lenticular, or fly's eye, lens methods.

Some may have noted that the systems just described are similar geometrically to the tried and true parallax barrier autostereoscopic systems.

The great advantage of the light line technique over the former system is its light transmission efficiency and resulting brightness. A slit mask works by blocking out light. Typically, a slit width less than one-tenth the width of the space between slits is used. As a result, 90 percent or more of the light is blocked. The result is either a dim display or a very power-hungry display that uses a bright light source or display surface to overcome the light loss at the slit mask. Combinations of wider slits and thinner pixels have been suggested; these block out less light, but the resulting amount of illumination is very uneven across the viewing zone.

A light line display retains the simplicity and freedom from optical aberrations inherent in the slit mask but avoids light loss problems either by directing light with high efficiency into individual light lines or by generating light at those lines. The result is much less wasted light and a much brighter display with lower power consumption. These are very important considerations in many types of displays, such as cockpit displays, that must have low power consumption yet be bright enough not to get washed out in direct sunlight.

The parallax illumination method also has several advantages over the lenticular lens systems. These include the 3D/2D switch capability mentioned earlier, the lack of optical aberrations associated with lenses of any type, and the capability of providing full-resolution images to each viewing zone. Parallax illumination methods also allow a greater number of usable viewing zones in front of the display than lenticular lens displays.

9.7 A Stereoscopic Three-Dimensional Drawing Application

In a joint venture between DTI and North Carolina State University, an object-oriented 3D drawing application named 3-D Draw was developed to address some issues of human interface design for interactive stereo drawing applications [CARV91]. The system was designed to execute on a Macintosh II using the standard Apple operating system including the Quickdraw [APPL88] graphics routines provided by Apple.

Menus and other 2D operating system features are displayed on the autostereoscopic display but are sometimes uncomfortable to view due to loss of horizontal image resolution resulting from sampling alternate columns of pixels. This sampling problem could be corrected by a double-resolution display, which is already in the prototype stage.

The Sampling Problem

A 3D image on the stereoscopic display resides in two buffers, one for each eye. It may appear at first that neither image can be drawn into its buffer using standard Quickdraw ([CHER88], [CHER90a], [CHER90b], [PARR90a], [PARR90b], [APPL88]) (two-dimensional Macintosh drawing package) routines directly, since the object must first be sampled along alternate pixel columns. This, however, is not the case; objects are rendered directly using a vertical line "pattern." Drawing with patterns is supported in Quickdraw, plus the pattern fills are mapped based on screen coordinates rather than by a coordinate system local to the object. Hence, like patterns of two adjacent objects will match or "line up," and the pattern mapping will stay fixed relative to screen position even with movement of the drawn object. Therefore, a vertical line pattern can be created to match the vertical columns of pixels for each stereo buffer. Moreover, a standard drawing mode is employed such that only the "set" or black bits of the pattern are copied/drawn to a buffer, preserving any images already in the other interlaced buffer.

Since the autostereoscopic display sends only alternate pixel columns to each eye, steps must be taken to remedy the image consistency lost during sampling. For example, vertical lines that are only one point (pixel size) wide may be lost completely to one eye but be visible to the other. Also, any object with vertical elements one point in width may not appear complete to *both* eyes; for example, a circle with a one-point frame width may not appear as a closed loop at its left and right edges. To address the sampling problem, first note how the user can see only "odd" line widths: a drawing width of 1 produces a line of width 1 to only one eye, a drawing width of 2 produces a line of width 1 to both eyes, a width of 3 produces a line of width 1 to one eye and of width 3 (with the interior pixel column removed due to sampling) to the other, a width of 4 yields width 3 to each of the eyes, and so on.

This trend yields only lines having an odd screen width, with "interior" pixel columns removed by sampling. Hence, 3-D Draw supports only odd drawing widths, by actually drawing the image with a pen width equal to the user-selected size plus 1. The standard Macintosh drawing pen can be redefined to incorporate these various heights and widths; when the user selects a drawing pen size of n, the 3-D Draw pen for the object is defined to have the rectangular dimensions of $n + 1$ by n. Since n is constrained to be odd, an effective drawing line width of $n + 1$ will be even. The vertical pattern mapping will thus retain a screen-generated width of n for each eye. Therefore, the picket fence sampling method does not prevent accurate presentation of odd vertical line widths.

Text Handling

The 3-D Draw application does not support even vertical drawing widths, but "ghosting" can still be present on the display, especially in text or other Macintosh operating system attributes. A problem sometimes occurs with the drop-down menus. The menus appear as a shadowed box with a list of textual commands. The left edge of the shadow box is one point in width and is thus visible to only one eye. Enabled commands appear as normal text, whereas disabled commands appear as gray pattern-filled text. The gray pattern is similar to the vertical line pattern but with alternate "rows" of pixels shifted. The sampling can render text of this form difficult to read. Placing checkmarks beside enabled commands and leaving disabled commands as normal text is an alternative solution but is contrary to Macintosh application standards. Normal text also shows problems if any aspect of it has vertical portions of width 1. Using boldface type may correct this aspect, but boldfacing also shrinks the open areas, such as the "hole" in the letter *a*. Such "holes" may be of width 1 and thus be visible to only one eye. Rather than compensate for these remaining artifacts, the 3-D Draw implementation follows Macintosh application standards and leaves the solution to be resolved with the production of a full-resolution monitor (nonsampled).

A screen alignment indicator was provided in the upper left region of the window so that the user could tell when the head was in the correct position. This feature may be toggled on and off as necessary. The indicator is a stereo image consisting of an arrow pointing left in the left frame buffer, and an arrow pointing right in the right frame buffer. The user may alternately close his or her left and right eyes to see the appropriate arrow.

The time-parallel presentation of images with this display architecture eliminates some problems caused by slow hardware and field-sequential stereo. Operating system features, however, must be added to handle presentation of true 3D objects, and multiple stereo windowing and pattern generation methods must be added to facilitate 3D text generation.

9.8 Conclusions

Parallax illumination is a very simple yet very versatile autostereoscopic technique that avoids many of the pitfalls associated with previously described autostereoscopic technologies. Its inherent strengths include brightness, lack of optical distortions, capability of compact embodiment in a flat-panel system, lookaround capability without loss of resolution, potential for great ruggedness, and usability across a wide range of applications with a wide variety of input.

Parallax illumination can, in the near future, lead to off-the-shelf flat-panel autostereoscopic displays that can be viewed as easily as conventional 2D displays, without the viewing aids (glasses), head position restrictions, or distortions normally associated with stereo. Such displays can be expected to eventually rival the venerable CRT display in terms of resolution and color. The availability of such displays, combined with increased use of 3D software, could cause stereo to become a widely used feature in many computer graphics and closed-circuit TV applications.

10

Chromostereoscopy

Richard A. Steenblik

10.1 Introduction

As the name implies, *chromostereoscopy* is a technique for converting color into stereoscopic depth. The process is unique because the encoding of depth is accomplished in a single image. Unlike conventional stereoscopic processes, a chromostereoscopic depth-encoded image retains all of its two-dimensional usefulness. It can be viewed with the unaided eye as a normal 2D image or through chromostereoscopic glasses as a stereoscopic image. Some versions of these glasses even allow the viewer to actively control the amount of depth seen in an image. The process is compatible with a wide range of display media, including photographic, printed, video, computer graphic, and laser-generated color images.

Although special optics are generally required to produce a useful depth separation, a small chromostereoscopic effect can sometimes be seen with the unaided eye as the result of common physiological and psychological conditions.

10.2 Physiological and Psychological Chromostereoscopy

The *physiological chromostereoscopic* effect can be observed in a two-dimensional image having bright red and blue areas scattered against a dark background. Most people will perceive the red areas as lying in a plane that is slightly closer than the plane of the blue areas. The effect appears to be due to the inherent chromatic dispersion of the eye acting on the off-axis component of the light entering it [SUND72]. (See Chapter 4 for discussion of this effect.)

The physiological chromostereoscopic effect can be enhanced by partially blocking the light to the eyes' lenses. Shadowing the half of each lens near the nose makes the outer halves of the lenses act more like prisms. This is quite easy to accomplish. First, make the "Boy Scout sign" with your right hand by extending the index, middle, and ring fingers and folding the thumb and little finger into the palm. Then lay the middle finger along the bridge of your nose so that the ring and index fingers partially obscure the view. The ring finger will block the right side of the left eye, and the index finger will block the left side of the right eye. Now look at the left- or right-eye view of Colorplate 10.1. To find the best viewing distance, some experimentation is required. Most people will be able to see a small depth separation between the red and blue objects, with the red in the foreground.

Our natural experience reinforces this color-dependent depth ordering. Red objects, such as fire, flowers, meat, and fruit are usually small and nearby. Even the red sun at sunset is often perceived as being near at hand. Large and distant objects, such as the sky, mountains, and the sea are usually blue. Artists and advertisers have used this *psychological chromostereoscopic* effect to great advantage.

Although they are both interesting phenomena, physiological and psychological chromostereoscopy can create only a very limited perception of depth. To achieve a more useful depth effect, it is necessary to resort to optical devices external to the eye.

10.3 Single-Prism Chromostereoscopic Optics

The simplest optical system for producing an enhanced chromostereoscopic depth effect is based on single-prism optics, so called because a single prism is placed in the line of sight of each eye [KISH65]. Experimentation with single-prism optics quickly proves both tantalizing and disappointing. An enhanced chromostereoscopic effect can be seen, but the cost of this effect may be instant, severe headache from the excessive visual accommodation required to fuse the image. Other negative side effects are vertigo, nausea, and visual disorientation.

Chromostereoscopic Depth-Ordering Modes

Assuming a direct relationship between the spectral position of a color and its perceived depth, there are two fundamental depth orderings available in chromostereoscopy. One is red to the foreground, or RF, in which red appears closest to the viewer, blue the most distant, and the remaining colors in between according to their spectral positions. The other case is blue to the foreground, or BF, in which blue appears closest and red most distant. The optical arrangement for producing the BF mode is shown in Figure 10.1.

Limitations of the BF-Mode Single-Prism Optics

Single-prism optics acts on an image in two important ways: it causes an overall deviation in the line of sight to the image, and it performs a chromatic dispersion of the colors in the image. Of these effects, only the latter is desirable. As Figure 10.1 illustrates, the deviation of the line of sight causes the viewer to turn his or her eyes inward in order to form a fused image. In this figure a circle indicates the location of the object, and a square signifies the position of the image component. The letters

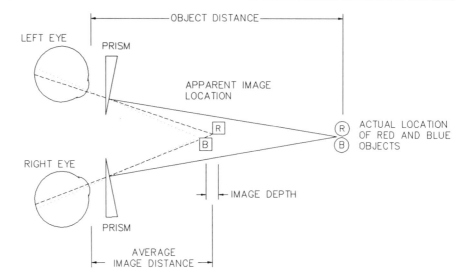

Figure 10.1
BF-mode
single-prism
chromostereo-
scopic
glasses.

in these shapes signify the colors red and blue. As this figure shows, the resulting image appears closer than the object because the lines of sight converge at a point closer in space. However, the focus of the eyes must remain at the distance of the original object. It is this disparity between the convergence distance and the focal distance that causes the aforementioned visual distress. Head movements are particularly discomforting. The image appears to move in an abnormal fashion because it appears closer than it actually is, causing the balance and position data sent to the brain from the eyes and the inner ear to conflict. The resulting sensation is akin to debilitating seasickness. Queasiness may persist for some hours after the glasses are removed.

The desirable effect of the prisms, dispersion of the colors in the image, is what creates the perception of depth. The lines of sight to each of the colors are different, so the convergence points for each color are different. It is important to note that the perceived image depth is not as great as it would be if the average image distance were equal to the object distance. The reason for this is geometric. As the image moves further away from the viewer, a given angular difference between red and blue will amount to a much larger perceived difference in depth. The image depth is not materially improved by increasing the prism angle. Increasing the prism angle will increase the angular separation between the red and the blue images, but it will also force the whole image to a point even closer to the viewer. The increased angular separation then acts over a smaller distance, effectively neutralizing the gain.

Limitations of the RF-Mode Single-Prism Optics

Reversing the orientation of the prisms also reverses the depth ordering, resulting in the RF mode, as shown in Figure 10.2. Unfortunately, this does

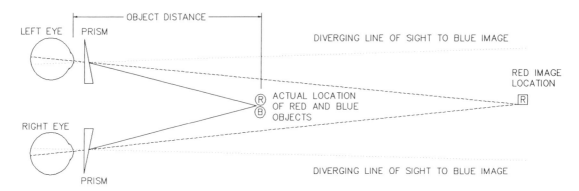

Figure 10.2 RF-mode single-prism chromostereoscopic glasses.

not improve matters. In this case the lines of sight to the image diverge from the lines of sight to the object. Except for the case of very small prism angles and very short object distances, the lines of sight to some or all colors will diverge past parallelism. The BF mode was bad enough, requiring the viewer to go cross-eyed, but this is worse. To see a fused image, the viewer must go *wall-eyed*. This skill is not in the visual repertoire of most people. Those colors that are not divergent will cause visual accommodation eyestrain in a similar manner as before.

Other Single-Prism Geometries

Other orientations of single-prism optics can be considered, but they solve none of the problems inherent in the two previous arrangements. If the prisms are placed in any parallel orientation, so that their apexes are pointed in the same direction, no depth is perceived. A single prism placed in front of one eye in either orientation yields approximately the same effect as the use of two prisms, one in front of each eye, each having half the single prism's apex angle.

10.4 The Superchromatic Prism

The problems and limitations of single-prism optics stem from the disparate visual convergence distances of the actual object and its image. These problems can be alleviated by an optical system that centers the image space on the object plane while retaining the chromatic dispersion function. In this scheme a single color is chosen to appear at the same distance as the object plane. A median color, such as yellow, is usually chosen. Other colors in the image are made to appear in front of or behind the chosen color in accordance with their spectral relationship to it. This places the average image distance at approximately the same distance as the object. Visual fatigue is minimized by reducing the difference between the convergence distance and the focal distance.

This effect can be achieved through a double-prism arrangement called a *superchromatic prism*. Superchromatic optics are the functional opposite of achromatic optics. An achromatic optical system is designed to achieve a spectrally invariant refractive function by *minimizing* chromatic dispersion. Superchromatic prisms are designed to *maximize* chromatic dispersion and to minimize other refractive effects.

A superchromatic prism consists of two prisms placed face to face with their bases pointing in opposite directions. One prism is a high-dispersion prism; it performs a chromatic dispersion function and imparts an angular deviation to the line of sight. As illustrated earlier, a single prism cannot avoid changing the line of sight. The second prism counteracts this angular deviation for the chosen color by applying an angular deviation equal in magnitude but opposite in direction. The chosen color exits along a path that is slightly laterally displaced from, but parallel to, the entering ray, as shown in Figure 10.3. The second prism is a low-dispersion prism. It performs a small chromatic dispersion function on the image in a sense opposite to the first prism. The net effect is that the lines of sight to the chosen color converge on the object while the lines of sight to the other colors converge at points in front of or behind the object plane.

The high-dispersion prism is made from a material that has a high coefficient of chromatic dispersion, meaning that the refractive index of the material is strongly wavelength-dependent. The low-dispersion prism is made from a material having a low coefficient of chromatic dispersion. The refractive index for this material is nearly invariant across the visible spectrum.

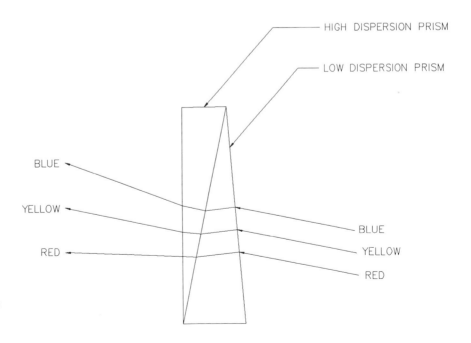

Figure 10.3
The superchromatic prism.

A variety of optical materials are suitable for creating these prisms, including glasses, plastics, and liquids. The prism angles, and therefore the thickness and weight of the optics, are minimized if the difference in chromatic dispersion between the two materials is chosen to be as large as possible. For this reason most superchromatic prisms fabricated to date have used optical liquids held in prism-shaped glass cells. The high-dispersion liquid is usually oil of cassia (cinnamon oil) or its synthetic counterpart, cinnamic aldehyde. These liquids have a chromatic dispersion unequaled by any glass or plastic optical material. The low-dispersion liquid is usually glycerine, which has an extremely low coefficient of chromatic dispersion.

Binary optics have also been used to simulate a high-dispersion prism having a low overall image deflection. Chromostereoscopic glasses fabricated in this manner hold great potential for low-cost mass production, which is discussed in greater detail in Section 10.8.

10.5 Superchromatic Glasses

BF-mode chromostereoscopic glasses can be constructed by arranging a pair of superchromatic prisms as shown in Figure 10.4. Comparison of this figure with Figures 10.1 and 10.2 demonstrates the improvement achieved through the use of superchromatic prisms. Glasses of this type can be designed to produce virtually any image depth desired.

The angles of the prisms can be chosen to center the image distance of any desired color at the object distance. In Figure 10.4 the image space is shown centered about the object plane, with a median color, such as yellow, centered at the object plane distance. This is generally the most desirable

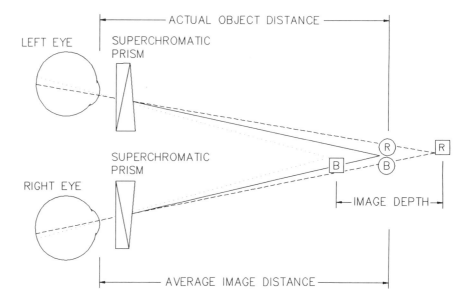

Figure 10.4
BF-mode superchromatic glasses.

image position, with some of the image lying in front of the object plane and some behind. For some applications it may be desirable to alter this relationship. Instead of placing the middle of the image space at the object plane, one may choose to place the background plane of the image at the object plane. This creates the impression that the image is floating in front of the display surface. For other applications it may be desirable to place the foreground plane of the image at the object plane. This yields the effect of looking into the display surface as though it were a window into a box. Centering the image space on the object plane combines both effects and minimizes the convergence versus focal distance disparity.

10.6 Conventional versus Chromostereoscopic Depth Encoding

The Conventional Approach

True stereoscopic processes generally share the basic functions of (1) encoding depth information into an image, (2) decoding the depth information to create left and right views, and (3) presenting the views selectively to the respective eyes.

The depth-encoding function generally requires that left and right views of the scene exist. These scenes are then individually coded in some manner, such as by color, time sequence, or optical polarization, so that they can be presented to the viewer as a combined image. When the image is viewed, the depth information is extracted by a depth-decoding system, such as color-coded or polarized glasses. The depth-decoding system isolates the views and presents them to the viewer's eyes. The chromostereoscopic process does not follow this process sequence.

Chromostereoscopic Depth Encoding

Chromostereoscopic depth encoding does not require the existence of left and right views of the image. Rather, the required data are a single two-dimensional image and specification of the desired depth position of each point in the scene. In the chromostereoscopic process this depth information is converted into color. In general, all regions sharing the same depth position will be given the same color.

Mapping depth into color is easily accomplished in created images, such as computer graphics and projected laser images. Indeed, the colors employed in these media are frequently arbitrary. In addition to artificially depth-encoded images, a surprising number of natural images, such as magazine photographs, contain chromostereoscopic encoded depth cues. These images can be viewed quite effectively as chromostereoscopic images.

Chromostereoscopic depth encoding involves applying a very simple set of rules to determine the color of each component of an image. It is convenient to consider the extremes of the visible spectrum to be red and blue.

Printer's inks, video monitors, entertainment lasers, and photographic pigments are generally bounded within this range. These are the primary media for the application of the chromostereoscopic process. Red and blue therefore define the extrema of the foreground and background positions in a chromostereoscopically encoded image. The other colors will generally lie at intermediate depth positions between red and blue.

As previously described for single-prism optics, there are two fundamental modes of mapping depth into color, RF and BF. Which mode is perceived in the depth-decoded image depends on the configuration of the decoding optics. The optical configuration shown in Figure 10.4 yields a BF-mode depth effect. The depth-decoding mode can be switched to RF by the simple rotation of each superchromatic prism 180° about a vertical axis through its center.

The mapping of color into depth strictly according to the spectral position of the color applies only to isolated colored regions presented on a black or very dark background. The use of other background colors can significantly alter the perceived depth relationship of the colors. The brightness of the background as well as its hue comes into play. In the extreme case, if the background is changed from black to white, the depth ordering will usually *reverse,* and the total depth in the image will be greatly reduced.

10.7 Tunable Depth Chromostereoscopic Optics

The unique properties of the superchromatic prism make it possible to create depth-decoding optics that allow the user to select both the decoding mode, RF or BF, and the maximum depth effect of the image. The user can smoothly scan through the entire range of depth from maximum RF, through zero depth, to maximum BF.

Tunable Chromatic Dispersion

The depth perceived in a chromostereoscopically encoded image is almost entirely dependent on the optical configuration of the decoding optics. To understand why this is true, it is necessary to examine the properties of the superchromatic prism in more detail.

The superchromatic prism can be considered an optical window that imparts a chromatically dependent angular shift to the light passing through it. This shift can be represented as a vector, having direction and magnitude. If the angle shift for red light is taken as a baseline, a dispersion vector can be found that indicates the angular shift of blue light away from this baseline. The dispersion vector points from the apex of the high-dispersion prism to its base, as shown in Figure 10.5.

If two identical superchromatic prisms are placed in series, the combination will have a dispersion vector that is the vector sum of the individual dispersion vectors. Note that only the horizontal component of the dispersion vector contributes to the perception of chromostereoscopic depth.

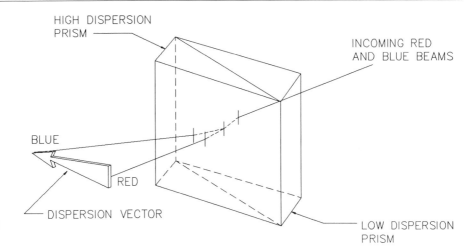

Figure 10.5
The dispersion vector.

The vertical component merely smears the colors in the image in a vertical direction. If a pair of superchromatic prisms placed in optical series are mechanically linked so that the vertical components of the dispersion vectors are always equal and opposite, then the horizontal component of the total dispersion vector can be varied at will by counterrotating the two prisms. The magnitude of the horizontal component can range from zero to two times the total dispersion of one of the superchromatic prisms. This is illustrated in Figure 10.6, where the horizontal arrow below each crossed pair of superchromatic prisms represents the dispersion vector resulting from that orientation.

Tunable-Depth Glasses

Tunable-depth glasses are created by placing two superchromatic prisms in front of each of the user's eyes and mechanically linking them for counterrotation. The arrangement shown schematically in Figure 10.7 could create the perception of intermediate depth in the RF mode. The user can choose any image depth within the available range by simply rotating the superchromatic prisms to the proper position.

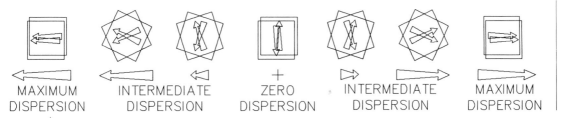

Figure 10.6 Dispersion vectors of cross-oriented superchromatic prism pairs.

Figure 10.7
Tunable-depth
chromostereo-
scopic
glasses.

10.8 Binary Optic Glasses

Until recently chromostereoscopic glasses were handmade by the author using optical liquids in glass cells. In 1990 the author and a colleague, Fred Lauter, began to explore the possibility of using the new field of binary optics to create inexpensive, mass-producible chromostereoscopic glasses. Work in conjunction with Gary Swanson and Margaret Stern, of Wilfrid Veldkamp's binary optics group of MIT Lincoln Laboratories, has led to the design of a binary optic that can closely emulate the performance of the liquid optics. This binary optic can be reproduced in plastic film through proprietary techniques. It packs the chromatic dispersion of a half-inch-thick liquid superchromatic prism into a plastic film only a few thousandths of an inch thick.

10.9 Unique Features of the Chromostereoscopic Process

Chromostereoscopic glasses perform their function in an entirely different manner from conventional stereoscopic image depth-decoding devices. Conventional stereoscopic processes selectively pass one of the two coded images to each eye by blocking the unwanted image from view. Chromostereoscopic glasses pass all of the image information.

Instead of blocking part of the image, the glasses selectively operate on it to *create* left and right views. The left and right views do not exist in the original image; they exist only after the image passes through the superchromatic prisms.

Most conventional stereoscopic processes are sensitive to the head orientation of the viewer. The stereoscopic effect is usually destroyed if the head angle goes below 15° from vertical. The chromostereoscopic process is completely insensitive to head position because the left and right views are carried with the glasses. Since the glasses move with the observer's head, they are never misaligned.

Advantages

By mapping the dimension of depth to the dimension of color, the chromostereoscopic process accomplishes depth encoding in a single image. A chromostereoscopically encoded image retains its two-dimensional usefulness since it does not consist of overlapping left and right views. It can be viewed with the unaided eye as a normal two-dimensional image or through chromostereoscopic glasses as a stereoscopic image.

Chromostereoscopic depth encoding is considerably simpler in many cases than conventional depth encoding. This is particularly evident with the projected images employed in laser shows. In this medium, depth encoding consists merely of choosing the appropriate color laser to project an image at the desired depth. Chromostereoscopic depth encoding of CRT-displayed computer graphics is straightforward for simple images but may be difficult for complex images requiring precise location of image components in the depth dimension.

The chromatic encoding of depth information in an image establishes the relative depth relationships of the parts of the image. The perception of depth is determined by the optical configuration of the glasses, which may be designed to exhibit virtually any image depth in either the RF or BF mode. Tunable-depth chromostereoscopic glasses allow the user to actively control the perception of depth.

Limitations

The chromostereoscopic process works best with pure colors. Spectrally narrow-band colors, such as those created by lasers, yield crisp, sharp images. Broad-spectrum colors, typical of the reflection spectra of some printing inks and the emission spectra of certain CRT phosphors, may produce fuzzy images.

Spectrally pure colors are preferable to composite colors. A small yellow dot on a CRT will break up into two dots, one red and one green. A sufficiently large region of composite color will retain its integrity, appearing at approximately the same depth position as if it were a pure color ([STEE87], [HODG88b]).

The use of color to encode depth information means that color is no longer arbitrary. The common technique of using color to highlight an image component would have the effect of changing its perceived depth. This can be controlled to some degree through judicious use of the background colors employed in the image.

There are obvious problems in applying the encoding scheme to images of people and real objects. We have strong expectations of the proper natural colors of the world around us, and arbitrary colors applied to such images can be disorienting.

10.10 Applications

Computer Graphics and Other CRT-based Images

CRTs produce all of their colors by combining various intensities of red, blue, and green phosphor emissions. The colors emitted by the phosphors differ in their spectral purity. Red phosphors tend to produce the most monochromatic light, green exhibits a somewhat broader spectrum, and blue is broadest of the three. The result is that red images appear crisp and distinct when viewed with chromostereoscopic glasses, green appears slightly fuzzy, and blue images are less distinct than green. Composite colors such as yellow, made by the combination of red and green, may break up into their component parts, depending on factors such as the size of the image, the color of the background, and the geometry of the image.

Hodges and McAllister [HODG88b] investigated the appearance of a wide range of CRT object and background colors as viewed through RF chromostereoscopic glasses. They concluded that red, yellow, green, and blue produced unambiguous depth information, and other composite colors did not. Both the background color and the colors of overlapping objects affected the perceived depth, sometimes in an unexpected manner. Overlapping objects of spectrally adjacent colors, such as blue and green or red and yellow, resulted in ambiguous or shifting relative depths. Unambiguous depth was observed for the combinations of red on blue, red on green, yellow on blue, and yellow on green. Other combinations produced inverted depth. A small red disk placed on top of a larger yellow disk placed in turn on top of a green disk yielded the impression of looking down a yellow cylinder with a red bottom in front of a green background.

In spite of these limitations, certain computer-generated images may be well suited to the application of the chromostereoscopic process. Images that may lend themselves to this process include CAD drawings, computer-processed medical images, weather radar, and air traffic control displays. Video games can be easily designed to make good use of this process; Mattel Electronics created an Intellivision video game in 1983 that was intended to be marketed with chromostereoscopic glasses. (Mattel sold their video game business before the game came to market, but the new owners of Intellivision made the game available under the name Hover Force. The game is not marketed with glasses.)

Hardcopy Printed Images

Excellent depth effects can be created with printed images. Unlike the case for a CRT, pure printing colors are not limited to red, green, and blue.

Many printing inks exhibit a very clean spectral reflectance and therefore produce a crisp image. As with all applications of chromostereoscopy, the greatest depth effect and most predictable results are obtained for images having a black or dark blue background. Colorplate 10.1 is a stereo pair illustrating what the viewer would see if he or she were wearing the glasses.

Laser Entertainment

Chromostereoscopy is exceptionally suited to the production of stereoscopic laser entertainment shows. The colors produced by lasers are spectrally pure and yield extremely crisp images. Control of the perceived depth is easily accomplished through control of the laser colors used. Several planetarium laser shows have applied this process on an experimental basis. The perceived image volume can be enormous, with images appearing as close as five feet and as distant as 140 feet. The perceived depth is not strictly limited to isolated color planes. Motion, change in image size, rotation, and other two-dimensional characteristics can be used to spread the depth over a continuous range, even with a limited palette of only four to seven colors.

Other Applications

There are a number of esoteric applications for this process. Recent research has shown that it can be applied to create stereoscopic images from binocular microscopes [STEE89]. By separating different-color objects into different depth planes, microscopic screening of different cells could be enhanced. Chromostereoscopy could also prove to be a useful tool in the teaching of astronomy. Star charts can be color-coded according to the red shift of each star, by distance, or by type, and then viewed with tunable-depth glasses to reveal previously unnoticed relationships. A similar approach might be applied to the teaching of anatomy.

10.11 Patent Status

The chromostereoscopic process is protected by a number of patents held by the Georgia Tech Research Corporation. Other patents are pending. Chromatek holds exclusive rights to the commercial application of the chromostereoscopic process, marketed under the name ChromaDepth.

11

The Oscillating-Mirror Technique
for Realizing True 3D

Lawrence D. Sher

11.1 Introduction

Oscillating-mirror displays provide spatial images in a display volume rather than flat images on a display surface. Like holography, they have depth rather than depth cues and have full parallax. Unlike holography, the user maintains interactive control of the spatial images. Images may be of the wireframe type or of the spatially distributed gray-scale type. Both kinds of images could be described as spatial distributions of luminosity (as opposed to gray-scale), and both are inherently transparent. A recent description of this technology, with emphasis in the field of protein chemistry, can be found in [SHER85].

11.2 History

The oscillating-mirror technique was first seriously investigated in the 1960s as the basis for a computer-driven display by Traub [TRAU68]. The large, lucid report that he produced is still an excellent introduction to the optical principles involved. Further Traub publications of the technique appeared in the late 1960s ([TRAU67a], [TRAU67b], [TRAU70]). In 1968 Rawson at Bell Labs investigated this technology in conjunction with a film loop for an image source and nonsinusoidal driving waveforms for his membrane mirror [RAWS68]. In 1969 he published a further discussion oriented more toward computer-driven image sources [RAWS69]. In 1970 Hobgood, a graduate student under the supervision of Brooks at the University of North Carolina, investigated some human factors issues [HOBG69].

In 1976 I built what may have been the first general-purpose, computer-driven display of this type; it was capable of showing up to 500 points using a membrane mirror, and it used only commercially available electronic hardware. In 1977 I upgraded the technology to 5,000 points by using a specially designed hardware addition to a PDP-11, and in 1978 I introduced the plate mirror. In the late 1970s, investigators at the Mayo Clinic, the University of

Utah, and the University of North Carolina built oscillating-mirror displays for internal use, primarily for raster-based medical applications.

In 1981 Genisco, under license from Bolt Beranek and Newman, brought out the first commercial display of this type. When this venture did not succeed and the license was returned to BBN, I undertook the design of a new, relatively small and inexpensive system, capable of wireframe drawings only. Several units were subsequently manufactured, sold, and used in military applications. Colorplate 11.1 illustrates this system. Many large-scale commercial exploitations of this technology have been contemplated, but so far none has been attempted.

In 1986 Getty and Huggins published a formal human factors study on the accuracy of perceived orientation and direction using this display technique [GETT86]. They concluded that there were "several significant advantages over stereo-pair and motion parallax displays, particularly in permitting the viewer to browse naturally and move about the displayed data" (p. 342). To my knowledge, no other formal studies have been done to explore the comparative merits of this kind of display technology.

11.3 Principle of Operation

The oscillating-mirror technology comprises a CRT-based monitor and an oscillating mirror in which the viewer sees the CRT's screen by reflection. Since the mirror is moving, so is the apparent position of the reflected screen. The apparent volume swept out by the moving reflected screen is known as the display volume. If one thinks of the moving reflected screen as intercepting successive, parallel layers of a spatial image, the display process can be thought of as successively displaying those layers, each containing an appropriate planar image fragment or slice.

When the CRT's face is dark, the volume apparently swept by the reflected screen (which henceforth will be called "the screen") appears to be a featureless dark void. When the screen is dark except for a point of light, then the swept volume appears to contain a point over which one has full electronic control of x, y, and brightness. By waiting for z to attain any desired value, one has a form of control over z as well. Because the movement is oscillatory at a sufficiently high rate, it is possible to repeat the process often enough that the eye sees not an occasional point but a nonflickering point fixed in space. To draw a line, it is necessary only to conceive of the line as a sequence of points, each of which is drawn in the same way, all drawn during a single cycle of the oscillation.

The means by which one can plot an illuminated path that looks like an arbitrarily branched structure, say a molecule, was conceived and put into practice in 1976 by me and later used in support of a patent [SHER78]. It can be reduced to a seemingly simplistic rule: draw what is possible when possible. For example, it is necessary to realize that the sweep of z caused by moving the mirror is very slow compared with the electronic speeds at which changes in x, y, and brightness are possible. Therefore, as z creeps (in tens of milliseconds) through its possible range, x and y leap (in 200 ns) to a suitable place, and the point briefly (400 ns) brightens to visibility; z continues to creep, and the point leaps and brightens again; and so on. Thus, whether a figure is branched becomes irrelevant, since lines are never drawn as lines but rather as points where they intersect successive z planes. (This brief description corresponds to "mode A," as described later in this chapter.)

To draw volume-filling data, such as a density function that has a single, scalar value at each lattice point in space, the data set is considered to be composed of multiple parallel planes. These planes are each drawn using a raster-scan technique, but the whole set of planes must be drawn during the time interval of a single mirror cycle. The result is a stack of separate planes in space that appears to map a spatial distribution of density into a spatial distribution of luminosity. (This description corresponds to "mode C," also described later.)

The foregoing paragraphs are a very simple explanation of a display technique that, in essence, is very simple. Important details will be expanded on in a series of categorical discussions.

11.4 Visual Perception

A Christmas tree filled with small lights is a 3D distribution of points of light. It is an easily understood model for the following thought experiment: Imagine that instead of many small light bulbs, there is only one. Every thirtieth of a second this light source sequentially visits all of the desired light bulb locations. If the bulb is dark while moving and bright when stopped, the appearance of the tree will be normal (minimal flicker). This idea, using one light source to mimic the presence of many, is the perceptual basis for the oscillating-mirror display. It is our good fortune that human vision is as willing to retain a 3D as a 2D afterimage.

The speed required for the single agile light source can be achieved with a CRT. (Currently available acousto-optically deflected lasers are too slow by about a factor of 10.) If one somehow oscillates a CRT along its axis, its phosphor screen repeatedly sweeps through a volume—the "display volume." Denoting screen coordinates by x and y and screen position in the volume by $z(t)$, then an image element could be written at any (x, y, z) by writing it at (x, y) at the correct time. The image element's appearance will be satisfactory—that is, it will appear stationary, of constant brightness, and of adequate contrast—if four conditions are met:

1. It is repeatedly rewritten at the same (x, y, z) location.

2. It is rewritten at regularly spaced time intervals that do not exceed about one-thirtieth of a second.

3. It is seen within a dark display volume.

4. It is not smeared in z by phosphor persistence.

A visual equivalent for oscillating a real CRT is oscillating its optically formed image. Optically speaking, the image can be real or virtual, but in practice, virtual is much easier since the time-varying optics can take a very simple form, as described next.

11.5 Optics

Mirrors of variable focal length, also called varifocal mirrors, were possibly first discussed by Muirhead [MUIR61]. Subsequent excellent discussions of the optics of varifocal mirrors can be found in [TRAU68] and [RAWS69]. The latter presents a more rigorous analysis than is given here, but it reaches the same final relationship by making a final simplification. (The final simplification there is done here at the beginning through the use of simplified definitions for p and q, which will be presented shortly.)

Figure 11.1 shows the simplest suitable optical arrangement, consisting of an oscillating plane mirror. Because of the equality of image and object distance, the "leverage" of the plane mirror is 2 (i.e., the image moves twice as far as the mirror). The size of the mirror determines the ease with which one can see the image, but it has no effect on the size of the image or its location. A plane mirror therefore must be large, must move through half of the desired image depth, and must oscillate at 30 Hz. This set of requirements is very difficult to meet.

Figure 11.2 shows a mirror that is forced to deform into alternately concave and convex shapes. With this design, assuming that the shapes can be kept optically useful throughout the motion, it is possible to use a very small amplitude of mirror motion yet maintain the desirably large amplitude of image motion. The price is constancy of magnification (always unity for the flat mirror).

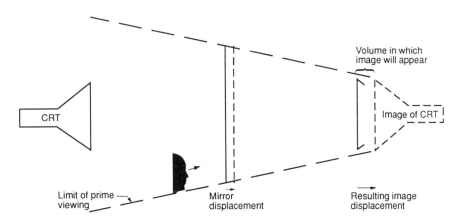

Figure 11.1
Oscillating plane mirror.

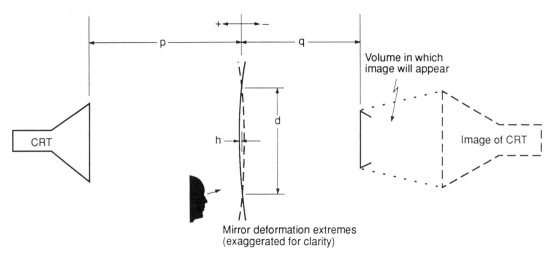

Figure 11.2 Oscillating deformable mirror.

Here the leverage (ratio of view volume depth to maximum mirror movement) is about 72, as can be easily derived. Let distances measured to the left of the mirror be positive and those to the right negative. In the following definitions of p and $q(t)$, distances are measured not to the center of the moving mirror but, as a simplification, to the center of the mirror before it is put into motion.

$$p = \text{object distance (fixed)}$$

$$q(t) = \text{image distance}$$

$$h(t) = \text{amplitude of mirror deflection at center}$$

Assuming that the mirror is a shallow spherical cap of radius $r(t)$, or at least osculates with such a sphere (i.e., it has the same first and second derivatives at the point of tangency—here the center), it is known that

$$\frac{1}{p} + \frac{1}{q(t)} = \frac{2}{r(t)} \tag{11.1}$$

Simple geometry shows that

$$r(t) = \frac{-1}{2h(t)}\left[\frac{d^2}{4} + h(t)^2\right] \tag{11.2}$$

(The minus sign arises because the sign convention of Figure 11.2 requires that $h(t)$ and $r(t)$ have opposite signs.)

For the typical case where $|h_{\max}| < 0.01d$

$$r(t) = \frac{-d^2}{8h(t)} \tag{11.3}$$

Therefore,

$$q(t) = \frac{1}{-16h(t)/d^2 - 1/p} \tag{11.4}$$

For sinusoidal time dependence, $h(t) = -|h_{max}| \sin \omega t$, where the minus sign has been added to make the final expression prettier. (The only other consequence of inserting a minus sign is a shifting of the phase of the sinusoid by half a period.) Then

$$q(t) = \frac{1}{(16|h_{max}|/d^2) \sin \omega t - 1/p} \tag{11.5}$$

For simplicity, it is convenient to let $A = 16|h_{max}|/d^2$ and $B = 1/p$, so that

$$q(t) = \frac{1}{A \sin \omega t - B} \tag{11.6}$$

For a plate mirror, one that overhangs its support, typical values are $|h_{max}| = 0.2$ cm, $d = 32.4$ cm (the supporting diameter), and $p = 67$ cm, so $A = 0.00305$ and $B = 0.0149$. Then

For the convex extreme, $\sin \omega t = -1$ and $q_{convex} = -55.7$ cm.
For the concave extreme, $\sin \omega t = 1$ and $q_{concave} = -84.4$ cm.

This calculation shows that while the center of the mirror moves through 4 mm, the image position moves through 287 mm, about 72 times as far. Or it shows that for an amplitude of mirror motion (measured 0 to peak, at the center) that is only 0.5 percent of the mirror's diameter, the extremes of image position are 28.7 cm apart. The image magnification, always given by $|q/p|$, is nonconstant, varying from 1.26 (concave case) to 0.83 (convex case). If the screen is rectangular, the display volume has the shape of the frustum of a rectangular pyramid (see Figure 11.2) whose apex is at the center of the mirror.

This discussion explicitly or implicitly reveals three potential problems:

Nonconstant magnification
Nonconstant velocity of the virtual image of the screen
A display volume that has an unusual shape

The nonconstant optical magnification has the property that it is known, with precision, in advance. Therefore, a compensatory electronic magnification on the CRT can be introduced. Its (time-varying) value is the reciprocal of the (time-varying) optical magnification, since one sees the product of the two magnifications.

The nonconstant velocity of the virtual image of the screen implies that at equal time intervals, the screen will have moved through unequal space intervals. Once again, electronics can compensate. If the controlling electronics does *not* use equal time intervals for showing successive image elements, but rather time intervals that are properly chosen, then the space intervals will be equal. Again, the problem is repetitive and well understood, so it is amenable to a repetitive solution. Other approaches that have been tried involve modification of the mirror-driving waveform ([RAWS68], [OBRI88]); however, these approaches can have dire acoustic consequences, which are discussed in Section 11.8.

The nonsimple shape of the display volume, a diminished rectangle in front linearly growing to become a magnified rectangle at the rear, is not a problem per se since its boundaries are not visible. But a problem remains: Because the larger the 3D image the better, one wants a simply shaped volume for ease of calculating the scaling factor, or a simple calculation that will use all of the volume. For its human factors qualities, the latter option is far better. Fortunately, an algorithmic solution to this calculation problem—how to scale and translate a desired 3D image so it is as big as possible within the confines of the frustum of a rectangular pyramid—is not excessively time-consuming. The calculation is of order N, where N is the number of points that define the wireframe or scatterplot image.

11.6 Mechanics

The circular mirror is forced into a concave-convex oscillation such that each point on the mirror's surface undergoes simple harmonic motion. A key design problem is that at every instant of time, the deformed mirror must be optically useful over its whole surface; that is, it must show a consistent reflected image regardless of the portion(s) of the reflective surface used. The other key design problem is keeping the acoustical output innocuous.

There are at least two viable methods for building a satisfactory mirror that is 30 to 50 cm in diameter. The first method, which uses a flexing membrane, is described by Traub ([TRAU68], [TRAU67a], [TRAU67b], [TRAU70]). The second method, which uses a flexing plate, is described by me [SHER78]. The membrane method uses a circular, metalized, plastic membrane (thickness 0.025 to 0.050 mm) held in a state of isotropic tension. It is deformed cyclically between convex and concave spherical caps by air pressure from an abutted loudspeaker. The metalization acts as the mirror.

The plate method uses a circular, metalized, plastic plate held in a normally tension-free state by a circular hinge concentric with, but smaller than, the plate, as in Figure 11.2. (A circular hinge permits only slight hinging action, which is limited by the elasticity of the materials, the result being slight concavity or slight convexity of the plate.) It is deformed cyclically between convex and concave paraboloids. Again, mechanical driving power comes from an abutted loudspeaker, and front-surface metalization of the plate acts as the mirror. Optical differences between spherical caps and paraboloids of such slight curvature are not significant for this intended use.

The overall diameter of the plate mirror used in the detailed example in the preceding section is 40 cm; it overhangs the 32.4 cm diameter of its supporting ring. A membrane mirror could be described by the same typical values, but since it cannot overhang its support, the overall diameter of the membrane mirror would be 32.4 cm. In other words, for a given image depth, the plate mirror has a bigger diameter for the same central amplitude of motion. Alternatively, if you start with a desired image depth and an overall mirror diameter, the plate mirror will have a significantly lower central amplitude of motion.

The only commercially available oscillating-mirror display uses a 40 cm plate mirror. The plate has weights mounted on its perimeter, the total mass of which is the primary means of making the plate mechanically resonant at 30 Hz. In its simple concave-convex mode of oscillation at 30 Hz, the required driving power is about 7 W.

Scaling up the plate mirror to larger diameters is limited by circumferential compressive stresses in the plate. These stresses can most easily be visualized by comparison with a muffin paper, whose concave shape is possible only when the paper is buckled, or pleated, circumferentially. Scaling up the membrane mirror to larger diameters is limited by the difficulty of driving it. Larger membranes must have more tension in order to suppress higher modes. The larger tension requires stronger membranes, which, in turn, require much more driving power. A concomitant of the greatly increased driving power will be increased acoustic noise. I know of no exploration of the practical upper size limits for plate or membrane mirrors.

In summary, a fair comparison of the nonoptical attributes of plate and membrane mirrors shows that plate mirrors are robust, quiet, easily driven, and difficult to build. Membrane mirrors are easier to build, relatively fragile, not as quiet (for equal diameters), and, for larger sizes, not as easily driven. Optically, both designs are capable of good to excellent performance. The actual performance is strongly dependent on details of the design and care in fabrication.

11.7 Synchronization

Since each image element must be refreshed by the CRT at a fixed (x, y, z) location, and since z is a function of time, there must be a method whereby the CRT "knows" when to light up (x, y) so that it will appear at the correct z. There are two basic methods for achieving this synchronization.

In the *open-loop* method, the CRT and the mirror are driven by a common source, so that frequency synchronization of their 30 Hz activities is guaranteed. Phase synchronization, however, is much trickier, since although there is essentially no delay between command and response at the CRT, there is a significant delay—possibly variable also—at the mirror. Nevertheless, once correctly adjusted (e.g., using a visual test pattern), such open-loop schemes can be satisfactory.

In the *closed-loop* technique, the mirror position is somehow sensed at least once per cycle, and this information is used to control the timing of all other events. Technically more difficult but potentially more accurate, this scheme has the disadvantage of requiring additional hardware. In the commercial implementation of a plate mirror design in 1980, a phase reference was successfully derived from the mirror by using an optical device to detect the passage of black-white transition on the edge of the mirror.

11.8 Acoustics

The mirror must attain optically useful shapes. Although this movement could be achieved in a vacuum or in an acoustically shielded enclosure, all mirrors I know of have operated in air and have been driven by air pressure. The upside is simplicity. The downside is the inadvertent production of acoustic noise levels ranging from unpleasant to intolerable.

The ear is notably insensitive to frequencies at or below about 30 Hz, but a frequency as low as 40 Hz is significantly more perceptible. Not only is the ear more able to discern this higher frequency, but the mirror system is more able to produce it! Since the velocity of sound in air is about 340 m/s, the wavelength at 30 Hz is about 1,130 cm, and at 40 Hz it is about 850 cm. A 40 cm mirror is therefore about 3.5 percent of the 30 Hz wavelength in air, and it is 4.7 percent of the 40 Hz wavelength. Since the efficiency of the mirror for acting as a loudspeaker varies with this percentage, using the higher frequency as the fundamental mirror-driving frequency, or as a harmonic, is hazardous for acoustic comfort.

Attempts to optimize some property of the mirror system, such as the constancy of image velocity, lead to attempts at carefully selecting just enough harmonics to approximate some desired driving waveform. All known attempts at this kind of waveform tailoring have led to acoustic disaster. Yet another example is manifest in a recent patent [OBRI88]. Therefore, at least for mirrors operating in air, one should keep the mirror motion as purely sinusoidal, at or below about 30 Hz, as possible. At 25 Hz, however, visual flicker may become troublesome. The compromise between acoustical and optical problems is a frequency window, perhaps extending from 25 to 35 Hz, where mirror operation (in air) is adequately free of both flicker and acoustic noise.

For plate mirrors, two design factors contribute to quietness. First, the mirror is designed as a mechanically resonant structure, with its fundamental resonant frequency (in the mode of interest—one circular node) at the desired frequency of operation. This arrangement is energy-efficient and has clean sinusoidal behavior. Second, since the circular plate mirror vibrates with one concentric circular node, the edge of the front surface recedes as the center advances, and vice versa. This behavior reduces the acoustic radiation efficiency of the plate.

11.9 CRT Technology

Assuming that the image source is a CRT (other possibilities exist, such as a film loop or laser device), there are two key requirements. First, the phosphor must have an extremely short persistence, since any afterglow will cause a smear of the image in the z-direction. This constraint precludes the building of a satisfactory oscillating-mirror display from commonly available components. I have found the uncommon rare-earth phosphor P-46 to be very satisfactory and the very common P-31 and P-4 phosphors to be completely unusable; for the latter phosphors, the observed persistences are too long by at least a factor of 20, despite tabulated properties suggesting differently. In the mirror, the phosphor screen appears to have a time-varying velocity that peaks at about 3,000 cm/s. (It is interesting that 3,000 cm/s divided by 72 is close to 1 mile per hour, the peak velocity of the center of the mirror.) Thus, an afterglow visible for 1 ms for a single plotted point will appear in the mirror as the desired point plus a 3 cm cometlike tail.

The second key requirement is that the CRT have electrostatic deflection rather than the more common magnetic deflection. The simple reason is beam agility. Electrostatic deflection is about ten times faster than magnetic deflection, and this factor of 10 exactly spans the boundary between unsatisfactory ($\sim 6 \, \mu s$) and satisfactory ($\sim 0.6 \, \mu s$) performance in point plotting. Drawing rasters is a somewhat more complex case, and it is possible that magnetic deflection is satisfactory for this purpose.

All oscillating-mirror displays to date have been monochrome. Color should be possible, but I know of no work on this potential enhancement. For a color display to be CRT-based, a suitable CRT may need to be designed.

11.10 Modes of Operation

The action to control a CRT's x- or y-axis is always a "plot" or a "sweep." With a plot, the x position, for example, is set to a specific value, which will be maintained until further notice. With a sweep, the x position is ramped, usually linearly, from one value to another. Since a raw CRT has two spatial axes, and each can operate in either of these two modes, there are four possibilities. Eliminating plot x/sweep y, which is simply a symmetrical variant of sweep x/plot y, there are three essentially distinct graphics modes, and all are in common use in 2D displays:

Plot x/plot y, a vector or directed beam display device
Sweep x/plot y, a time-base oscilloscope
Sweep x/sweep y, a raster display device

The moving mirror creates the appearance of another spatial axis, z, which is swept in an oscillatory manner. A direct extension of the foregoing categories of possible graphics modes, then, is (where b is brightness):

> *Mode A:* Plot x/plot y/sweep z (data in x, y, and b)
> *Mode B:* Sweep x/plot y/sweep z (data in y and b)
> *Mode C:* Sweep x/sweep y/sweep z (data in b)

Mode C has an interesting variant that has never been built but is potentially useful. For conventional 2D presentations, there is little reason for the sweeps in x and y to have anything but simple, linear, repetitive, sawtooth waveforms. But for 3D presentations, every individual swept line could start and stop at any (x, y) location, and the brightness could be modified along the length of the line. With suitable software support, this more generalized sweep would enable the drawing of many pixel-based, nonparallel, planar images in the display volume during one refresh period. Since this multiplicity of nonparallel 2D rasters could not reasonably be described as a 3D raster, there is a gap here in our nomenclature. A possible patch might be to modify the description of mode C and add a mode D:

> *Mode C:* Sweep x/sweep y/sweep z (x and y sweeps are sawtooth)
> *Mode D:* Sweep x/sweep y/sweep z (x and y sweeps are data-dependent)

In a sense, all modes are special cases of mode D. Mode D would be useful for applications that generate multiple nonparallel 2D rasters sharing a defined volume. Ultrasonics is an important, speculative instance.

The graphical result of mode B is best thought of as a horizontal plane that is deformed vertically into "mountains." (Note the similarity to a time-base oscilloscope in which a horizontal line is deformed vertically.) Mode A can mimic mode B by simply plotting sequential x-values to create the digital equivalent of a sweep. The graphical result of mode D is unique. One would see multiple 2D gray-scale images intersecting in a dark and otherwise empty void. Very little developmental work has taken place on mode B and none on mode D. Modes B and D will not be discussed further here.

One distinction between modes A and C is illustrated in Figure 11.3. The essential idea is that a volume element at (x, y, z) can be lit to a specified brightness by one of two methods. As the diagram makes clear, mode C could entail a lot of waiting. Waiting is anathema to a refresh-type display, since there is a fixed interval of about 33 ms in which to draw the image. Therefore, mode C should be used only when there are image elements at most (and usually all) of the possible (x, y, z) locations. Otherwise, mode A should be used.

In summary, modes A and C have complementary strengths and weaknesses. Mode A can provide excellent spatial resolution for a sparse image, whereas mode C overcomes the sparseness limitation at the expense of

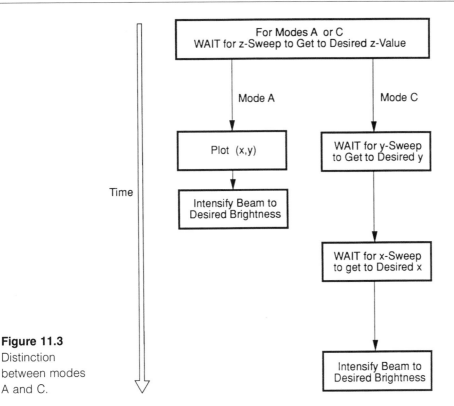

Figure 11.3
Distinction
between modes
A and C.

spatial resolution. Line drawings should be shown in mode A, and cloudlike images naturally lend themselves to presentation in mode C. This distinction, however, does not address the question of whether a given data set of the form $d(x, y, z)$ (e.g., a density function) will be easier to understand if it is somehow plotted in mode A. This question is addressed later in this section.

Use of Mode A

In mode A the CRT acts like a plotter rather than a raster device. As z is swept, a sequence of (x, y) values is plotted on the CRT. Since this plotting sequence is the essence of creating a mode A picture, and since there is only a finite time in which to complete it, the plotting must take place as rapidly as the hardware can accommodate. The set of points to be plotted is known as the display list. It contains only (x, y, b) triplets, the z values being implied by the ordering of the list.

Software support for mode A must deal with the problem of creating the display list from the logical description of the desired image (e.g., the vector list). That problem is key to the practical employment of mode A. The algorithmic problem to be solved is easiest to understand with a 2D analogy: How can a conventional 2D raster display show an arbitrary line drawing to best cosmetic advantage under the restriction that each raster line can have only one pixel illuminated? This problem appears to have no ideal solution, although there are many possible adequate solutions. As a practical matter, the most desirable solutions are the ones requiring minimum computation time.

If the display list contains all points that the hardware can plot at full speed (about one point every 600 ns), then the z intervals between points will be variable. That means that the calculations must take into account these variations. Alternatively, the display list can contain a subset of the list so that the z intervals are constant. The algorithmic calculations are eased, but at the expense of discarding some of the possible points. The latter plan has been found to be superior. The sacrifice is that approximately 40 percent of the possible points are not displayable, but the calculations are faster, and the point loss tends to occur toward the volume's front and back. Because most wireframe drawings have maximum complexity toward the center, the loss tends to go unnoticed.

For a mode A display device the number of points is the figure of merit, unlike the case for directed-beam displays, which are based on the number or lengths of long and short vectors. Since, by observation, 10 points makes 1 cm of solid vector (in the display volume, not on the CRT), 16,000 points potentially allows 1,600 cm of cosmetically unblemished line. Our visual trials show this length of line to be sufficient for many applications (modal analysis, scatterplots of 5,000–10,000 points, proteins and nucleic acid molecules, weapons trajectories, air traffic control, terrain, etc.).

In assessing the adequacy of 16,000 points as a drawing resource, one must note that the guide of 10 points/centimeter is very soft. The number of points can be reduced to 5 or even fewer, depending on the image, with satisfactory results. Thus, the actual length of line available for drawing could exceed 3,000 cm (100 feet). Line figures complex enough to require this length of line may approach a state of "steel wool," wherein detailed information content is effectively lost. Therefore, on the basis of the amount of line that can be drawn at one time, current technical capabilities are somewhat matched to the needs of mode A.

If mode A is pressed into the service of drawing surfaces, the number of available points may be found to be inadequate, but this assessment depends on how the surface is oriented, how much area is involved, and what resolution is sought. Drawing volumes (i.e., voxels) with the points of mode A is ill-advised; the digital bandwidth used for positioning control should be used instead for brightness control.

The traditional high-level description of a line figure is a list of nodes and connections, that is, a vector list. The foregoing description of the highly unusual way in which an oscillating-mirror display plots lines suggests that it is difficult to use this technology, but such need not be the case. Prototype and commercial systems have built-in algorithms for converting one kind of description to the other. Therefore, they can be used like ordinary graphics devices. There is no difference in the form of data input, only in the form of the displayed output. This important feature means that old vector lists and old programs for manipulating molecules, for example, work directly with the new form of true 3D output. The primary problem is usually one of formatting; a vector list may have to be changed in form but not in content.

The strength of mode A is its instant clarity. The image is crisp, high in contrast, and usually as easily understood as a physical model. It may even be more understandable than a physical model if opacity in the model interferes with perception of spatial relationships. Any mode A image weaknesses are probably related to the inherent limitations of the generic oscillating-mirror technique: the inability to naturally depict opacity; the restriction to drawing with white on black, and not vice versa; and, as commonly implemented, the inability to display objects further away than arm's length. Of course, color is still missing.

Use of Mode C

Mode C naturally lends itself to the display of three-dimensional scalar fields (e.g., electron density at each lattice point in a volume or 3D reconstructions from projections). The presentation provided by mode C is objective, free of the artifacts of slicing, and in a form admitting interaction with the viewer. Here "objective" implies that two viewers see the same 3D image, as opposed to seeing the same set of 2D images and then independently making mental constructs to 3D. "Free of the artifacts of slicing" implies that a cloud of variable density is displayed as such a cloud, not as a set of distinct 2D cross sections. "In a form admitting interaction" implies that the viewer can request and then see a revised presentation—for example, a spatial vignette excluding nonessential and possibly obscuring foreground and background imagery. This kind of interactivity is a major distinction between oscillating-mirror images, where one has the power of the computer for making real-time changes, and static holographic images, where current technology cannot make real-time changes.

These favorable attributes of mode C must be balanced against its nonfavorable ones. First, mode C images are not as well matched to the capabilities of commercially available electronics as are mode A images. Without delving too deeply into the details, suffice it to note that in the same time that TV provides one complete picture (two interlaced frames), an oscillating-mirror display must show many pictures. A quantitative analysis of sufficiency is not as easily formulated as it is for mode A, since mode C could be used for a wide variety of purposes. But the essence of the difficulty is that the ability to show many (say, thirty to fifty) relatively coarse 2D rasters within one-thirtieth of a second is not a well developed (or otherwise desirable) commercial technology.

The near-term outlook for an adequate number of sufficiently bright voxels in mode C (say, several million) is not favorable. Given user demand, industry might respond with variants of existing technology, such as multi-gun CRTs, but there are no signs of this at present. Lasers appear to offer more promise for this purpose.

Second, mode C images are unlike the visual images of the world we so easily understand. The images appear as a spatial distribution of luminosity. Along any given line of sight from each eye, there is a certain amount of radiant energy, which the brain must somehow interpret as a sum. This mental requirement, in fact, resembles reconstruction from projections! But with the added cues of head-motion-induced parallax and stereopsis, the brain is presented with quite a challenge. Human factors studies of such visual processes are sorely needed, so that optimal display formats can be designed. Until then, we must proceed on experience.

Experience with mode C has led to a few conclusions amid much remaining ignorance:

1. The sparser the image, the easier it is to see what remains. That observation leads to a major unanswered question and a good opportunity for future human factors studies: How should one make a mode C image sparse?

2. As noted, images are more easily understood if they are sparse and of high contrast. Thus, mode C images are more easily understood the more they resemble mode A images. This observation implies that there may be cases where a data set that lends itself naturally to a presentation in mode C might be shown to better advantage in mode A if it could somehow be converted. For example, a density function $d(x, y, z)$ might be shown in mode A as a set of isodensity surfaces, each found and then drawn with points or nets. No general rule can be formulated about which mode is best for such borderline cases. The ultimate choice should be made on the bases of practical constraints, the nature of the application, and visual trials.

3. Users must be able to interact with mode C images in order to best understand the informational content. Interaction could take many forms, such as vignetting, highlighting, performing some rotations, and brightness remapping and clipping.

4. Unless a mode C application is well defined, the partition of available voxels between x, y, and z should be programmable to best adapt to the needs of the application.

5. Subtle patterns may be discernible in mode C images that are not as easily found in any sequence of flat images.

6. The preservation of spatial context in the display of 3D scalar fields is one of the most valuable attributes of mode C images. In selected circumstances, its value might outweigh all disadvantages.

11.11 Using the Two Halves of the Mirror Cycle

The mirror spends as much time going from convex to concave as from concave to convex. The result is that a complete image can be drawn in either half (each sixtieth of a second) of this cycle. Furthermore, two images can be drawn by using both halves of the cycle, and since the full cycle takes one-thirtieth of a second, both images will be refreshed at the same 30 Hz rate. Thus, it is possible to draw two mode A or C images, apparently simultaneously.

It is tempting to imagine using this capability to show the same image in both halves of the refresh cycle, thereby doubling the effective refresh rate and allowing the mirror frequency to be lowered to 15 Hz. There are three serious problems with this idea: (1) The forestroke and backstroke of the mirror cycle would have to be exact symmetrical counterparts; (2) control of phase would need to be very precise; and, fatally, (3) those parts of the image near the front and near the back would effectively be refreshed at 15 Hz (two refreshes in rapid succession every fifteenth of a second), causing those parts of the image to flicker.

Productive usage of the two halves of the mirror cycle should focus on simultaneously showing two different mode A images, two raster-interlaced mode C images, or one mode A and one mode C image. For two simultaneous but different mode A or C images, exact registration is usually not critical, especially when the registration errors are likely to be in z, the least noticeable form of error. Showing one mode A image together with one mode C image would benefit those applications that seek to derive structure information from a 3D scalar field describing the structure. There are numerous such cases in biochemistry.

11.12 The Uses of One More Dimension

In fourteen years of experimenting with oscillating-mirror display technology, in exploring over thirty applications, and in showing this form of display to several thousand people and getting their reactions, I have found several ways in which this display technology appears to provide qualitative improvements over flat displays and stereo displays. I make no claim that one technology is better than another. I do claim that a large body of anecdotal evidence suggests that this new technology has certain domains of applicability in which its peculiar properties constitute a net advantage.

3D Data Can Be Shown Objectively

The oscillating-mirror display is capable of objectively representing a 3D image, where "objectively" means that, just as with a physical model, the appearance is not subject to interpretation; it simply is what it is. "Objectively" does not imply "realistically." In fact, realism for some subjects is undefined (e.g., molecules).

A conventional flat display uses depth cues that different people may interpret differently; the various interpretations are viewer-dependent. A stereo display uses another form of depth cue that works well for many but not all people, the latter group consisting of those with unknown minor to major visual troubles. A physical model, another form of 3D display, works well for all people, with the objective reality of the model allowing possible interpretation differences to be easily resolved.

Objectivity is a benefit when the viewer has no relevant prior knowledge, and it helps in a teaching or consultative situation. A more subtle point but one that is most important is that the viewer need not deal with the question, "What is it that I see?" but can go directly to the central question, "What is important in what I see?"

The User's Eye/Brain Pattern Recognizer Is Exploited

Human vision is the adaptive pattern recognizer par excellence. It even works in 3D. But often lacking is the means for getting the best synergism between this human ability and the computer's incomparable speed and memory. Oscillating-mirror displays can bring the two together in a useful way.

In true 3D, one can plot four variables at once, with brightness being the fourth. By adding motion, additional variables can be added. It is experimentally evident that the eye deals with the extra dimension perfectly well. In fact, were such display technology more widely available, it is likely that more experiments would be designed to take advantage of this way of seeing the data.

Representing the significant patterns found this way in a published form is not as difficult as it might seem. The usual problem is finding the patterns. Once found, it is not as difficult to find ways of showing the patterns to advantage in a flat publication medium.

Perishable Data May Become More Useful

Some data are perishable. If not understood quickly, perishable data become useless. An example is the state of a process controlled by a human operator. The operator must note the process parameters, decide if action is required, and take any such action in a timely way.

Process control with a human in the loop demands an understandable display of the current state variables. Good examples are piloting a vehicle, power plant control, and controlling a manufacturing operation. In all such cases it is necessary but not sufficient to have all information displayed or displayable. The important information must be easy to understand within a meaningful time interval, a virtue not always appreciated. In this case, the perhaps unexpected benefit of adding a third spatial dimension is that less time may be needed for understanding a set of data.

Some Difficult Data Can Be Displayed

Wind shear, fluid flow, electrostatic fields, spatial scatterplots, and spatial trajectories are all common forms of data that do not map well to flat media. The problem is dimensional mismatch. Conventional means of addressing the problem are to add color, depth cues, computer-generated motion, vignetting, and other enhancements. Particularly effective conventional strategies are to show only what can be shown, to show special cases, and to show parts of the data that can be rendered in a particularly satisfactory way; the implication is that anything else could have been shown. These methods are effective to various degrees, but they are attempts at circumventing the fundamental problem, dimensional mismatch.

With one more real dimension, the problem of dimensional mismatch is simply postponed, but it is an important postponement, since representing an entire class of data sets that were at the fringes of displayability may now be tractable, and in some cases, easy.

For example, with the extra dimension, oscillating-mirror displays can display the streamlines within any part of complex flow fields. Small points can march along the streamlines to indicate vector magnitude. The streamlines themselves can also change position slowly, with time slowed enough to facilitate understanding. The flow field might still be difficult to understand, but if so, the complexity would be inherent in the data, not in its display.

Another case of difficult-to-display data is a structure with internal movement. If computer-generated motion is used as a depth cue, then this display resource is not easily available for showing real movement. Here is a case where the benefit of one more spatial dimension may be a superior way of showing motion. A military application that has successfully used the oscillating-mirror display for this purpose is the spatial chase of modern weaponry.

11.13 Summary

The present state of oscillating-mirror display technology can produce moderately complex, monochrome, transparent, wireframe or raster-based images. There are many applications for which these properties are well suited and others for which these properties are not well suited. This addition to the means for computer-based display should be appreciated for what it is, not disparaged for what it is not. It provides three visually real dimensions that may be useful for many current and future applications.

12

Three-Dimensional Imaging through Alternating Pairs

Edwin R. Jones, Jr.

A. Porter McLaurin

12.1 Introduction

For many years artists and scientists alike have tried to develop methods for presenting images that recreate the three-dimensional space in which we live. Most of these techniques, from Wheatstone's first stereoscope to the present-day motion pictures seen with polarizing filters, are based on a common feature: the presentation of separate views to the separate eyes. Moreover, these stereoscopic displays all require two views derived from positions that are horizontally separated but vertically the same. We have developed a novel system without these requirements, one that alternately presents views with vertically displaced perspective to both eyes simultaneously [JONE84]. This technology, known as the VISIDEP or alternating-pair (AP) method of three-dimensional imaging, presents autostereoscopic images on a flat screen in a manner that does not restrict head movement in any way.

In this chapter we describe the basic AP system. We also detail some innovations that improve the image quality by suppressing the slight rocking motion inherent in the alternating-pair technique. Beyond these two basic areas, we describe some more recent applications of the alternating-pair technique.

12.2 Visual Memory

The AP process for three-dimensional imaging arose from research into the requirements for perception of depth in flat-surface images. Stromeyer and Psotka reported earlier ([JULE71], [STRO70]) that a certain "eidetiker" who looked at one view of a random-dot stereogram with one eye retained it in memory. When the other view was later shown to the other eye, the eidetiker recalled the first view and fused the two of them to perceive the stereoscopic image. Because random-dot stereograms are purely stereo-scopic and require a pair of images for interpretation, it was concluded that image retention and later comparison had taken place. This fact implies that one can view the two parts of a stereoscopic image at different times and still construct an image with depth. Such a finding means that the act of perception is a cognitive activity and not merely a stimulus response. This view is consonant with those stated by Rock [ROCK85] concerning the whole of perception.

That depth perception of time-displaced images is not limited to eidetikers has been known for some time. Ogle [OGLE63] reported in 1963 that when left- and right-eye information is alternately presented to the left and right eyes, depth perception is possible if the time interval between presentations is less than 100 ms. Three-dimensional television systems that incorporate this idea are already in the marketplace. They present left-eye/right-eye in-formation sequentially to the proper eye by the use of electro-optic shutters ([BUZA85], [ROES87], [STAR90], [LIPT89]). Fusion of the binocular im-ages in a short-term memory buffer has been suggested by Marr [MARR82] as the means by which the depth map of the image is stored. An interme-diate memory such as this accounts for the results observed by Ogle. His observations and those resulting from research with AP confirm Rock's idea of depth perception as a truly cognitive activity.

Because visual information to the two eyes is compared in a temporary memory and does not have to be received simultaneously, there is no reason that the stereoscopic information could not be obtained with a single eye, provided that the appropriate images are sequentially received at a rate appropriate for the memory buffer. Thus, by providing stereoscopically related images alternating at the proper frequency, the perception of depth can be realized. This phenomenon is central to the visual image depth enhancement process (VISIDEP) for the presentation of three-dimensional images by means of alternating pairs of stereoscopically related views.

Because the depth in alternating-pair images is generated from two slightly different viewpoints, motion of the observer's head produces no additional parallax. Thus, one cannot look around a solid object or see behind it, as can be done with holographic images. In this regard, the AP images are similar to those produced with stereoscopic glasses. In fact, lateral motion of the head creates an illusion that the three-dimensional scene is rotating to follow the head motion, an effect that is also observed with head motion when viewing traditional stereo scenes with glasses. However, the depth perception is quite distinct from that obtained with traditional stereoscopic systems in that one cannot roam through the depth of the scene at will by changing the convergence of the eyes. Instead, there is an immediate perception of depth that cannot be enhanced by changing the convergence of the eyes from the plane of the display screen. The perception of depth occurs even though both eyes are receiving essentially identical input. The perception is equally strong for a single eye; in fact, some observers have said that the depth perception obtained through AP for a single eye is even better than when both eyes are used.

Although it may seem that AP is a special case of the kinetic depth effect reported by Wallach and O'Connell [WALL53], the two effects are distinct. The kinetic depth effect is the three-dimensional interpretation of changing two-dimensional shadows of a rotating three-dimensional object, such as a wireframe model. The kinetic depth effect occurs when a shadow line undergoes simultaneous displacement and change of length. However, depth can be seen in random-dot stereograms presented via AP, even for scenes in which no length change occurs. AP may be more akin to monocular motion parallax [OKOS76] than to the kinetic depth effect.

12.3 Details of the Basic Process

The technique of alternating pairs consists of presenting on a two-dimensional screen a sequence of alternating views derived from different points of origin. These binocularly related images may be time-displaced and presented in their proper time sequence. The quality of the resulting 3D image perceived by a viewer depends on the rate at which the images alternate. When the rate is too low, the effect disappears, and the viewer merely sees a succession of flat images. When the rate is too high, the image appears to flatten out, and the perception of depth again disappears. In the limiting case of very high rate, both views appear to be on the screen at the same time, and there is no depth at all. However, at an intermediate rate of four to thirty alternations per second, the sensation of depth is readily apparent. We have obtained the best results for rates near ten changes per second.

When stereo-correlated images are alternated at a rate of about eight to fourteen changes per second, the perceived image is of a scene in depth, but it appears to have a slight, smooth, and continuous rocking motion, even though the images are being alternately switched off and on sharply. This effect is easily achieved with slides, motion pictures, or television in which the images from two sources A and B can be switched by a squarewave with equal dwell times (Figure 12.1). If the two images are alike in all ways except for their binocular parallax, and if the parallax is not too large, the rocking motion can be reduced to a point that it is barely perceptible. Of course, in the limit that the parallax is reduced to zero, there is really only one viewpoint, and the perception of depth disappears.

One reason for the effectiveness of this particular frequency range for image alternation is that it lies below the critical flicker rate so that the brain is able to consider both parts of the stereo pair and compare them to generate the depth image. Moreover, the optimal frequency rate of ten changes per second corresponds to the timing previously observed in the experiments of Ogle [OGLE63] and Julesz ([JULE81], [JULE83]).

Figure 12.1
Basic sequence
of output signals.

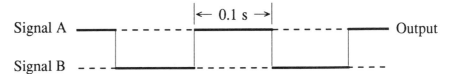

Signal A — Output

|← 0.1 s →|

Signal B

Most stereo systems are designed to mimic the interocular separation of human eyes by placing cameras so that their optic axes are horizontally aligned with a separation of 65 mm. When configurations of this type are used in the AP technique for normal scenes, the resulting on-screen image has too much motion arising from the change in viewpoint from one camera to the other as the images are alternated. If the effective offset between camera positions is reduced to 10 to 20 mm, the stability of the image is restored, and a more satisfactory effect in achieved.

There is, of course, the stringent requirement that the alternating images be carefully aligned on the screen. Registration errors due to translational misalignment on the screen or to rotational misalignment of the images about the optic axis are readily noticed and seriously degrade the perceived image. For best results, the alternating images must be identical in every respect except for the disparity due to their differing points of origin. Other image factors such as brightness, color balance, contrast, sharpness, and size must all be carefully controlled so that there are no apparent differences between the images. Failure to achieve the desired match does not reduce the apparent depth but does result in unwanted distractions that can render the images unacceptable.

Because of the unique feature that the time-displaced images can be received and interpreted with only one eye, the AP technique is not restricted to horizontal parallax. Tilting the head while watching the display produces no apparent loss of depth. Even for the extreme case of vertical image parallax, depth is still perceived. Surprisingly, we found the use of vertical parallax superior to the use of horizontal parallax. Hodges and McAllister [HODG85], working with computer-generated images, also concluded that vertical parallax produces a more satisfactory display.

In several instances we encountered images with horizontal parallax that were misinterpreted when viewed with the AP technique. For example, a stereoscopic pair of images of a complicated molecule was drawn by computer from x-ray data. When this stereo pair was viewed in a traditional stereoscopic manner (left image for left eye, right image for right eye), one saw three ringlike members oriented in a nearly vertical plane, with two other members of the molecule arranged dumbbell-like perpendicular to that plane. When the images were presented via the alternating-pair technique, however, the back ring appeared to move forward and was seen as projected toward the viewer. This was an obvious misinterpretation. When both figures were rotated by 90° to a position equivalent to vertical offset, the image appeared in the correct perspective, with the back ring assuming its proper position relative to the rest of the molecule. Thus, it seems that there is a difference in the manner in which the brain correlates the stereographic information, and that vertical parallax is superior to horizontal parallax for presenting depth information via the AP technique. We have encountered several other instances with computer-drawn images in which the depth confusion that occurred with horizontal parallax was eliminated when vertical parallax was used instead. This superiority of vertical parallax was especially noticeable when the technique was applied to stereo-related x-rays. Thus, by using sequential images with vertical parallax presented at a rate of approximately ten changes of viewpoint per second, the AP system presents a 3D image that can be viewed with the unaided eye. Because of the unique approach of this system, no special equipment is required for transmitting, broadcasting, receiving, recording, or playing back these 3D images in any video format.

12.4 Advances in Alternating-Pair Systems

The success of the initial AP method spurred efforts to reduce the rocking motion inherent in the process while preserving the illusion of depth. We discovered that significant reduction in the apparent motion can be achieved by having one of the images constantly present on the screen instead of completely alternating images as previously described. This effect has been achieved with an off-the-shelf special effects generator (SEG). With an SEG operating in the "mix" mode as an integral part of the system, we could alternate between a single view and a combination of both views [MCLA86], which produced a far more stable image. In this new configuration (Figure 12.2) the signal from source A becomes a constant baseline, and signal B is mixed with A at the rate of ten changes per second.

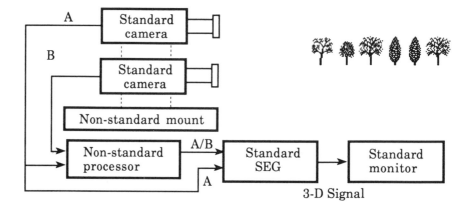

Figure 12.2
Typical
dual-camera
system.

Typically, the signals are mixed in a 50-50 proportion to achieve the desired effect. The nature of the mixer ensures a constant signal level output, eliminating luminance variations. The mixing technique substantially reduces the slight rocking motion mentioned by Hodges and McAllister [HODG85] yet still generates a depth-enhanced image with a realistic three-dimensional appearance.

An additional feature of the mixed-signal technique is that the mix can be controlled by the operator, permitting depth to be gradually faded in or out as the mix is varied from the extremes of no mixing (single source) to full alternation. With no admixture of signal B, there is no depth. As the amount of B increases, the depth gradually increases. As one moves toward full alternation between the signal sources, the maximum depth and the maximum motion occur. At roughly the 50-50 level, the apparent motion is significantly reduced with only a slight diminution of the depth effect. Consequently, greater disparities can be tolerated with the mixed-signal technique, a finding that may be valuable in further hardware development. The greater separation implies a greater tolerance for errors in alignment. Furthermore, the mixing technique is much more tolerant of mismatch in color, brightness, or contrast than the simple alternating method. Because one signal is always present, differences such as color mismatch between the cameras are averaged to an apparently smaller effect.

In normal binocular vision with parallel input, there is no perceived alternation between the left and right images. Instead, one sees a single steady image. Upon closer inspection, however, one notices that all of the world in view is not fused into a single image. For example, if you concentrate on a finger of your outstretched hand, you simultaneously see that objects in the background are not fused but appear as a double image. Similarly, if you focus on the background far away, your hand will appear as a double image. This observation, along with knowledge that it is possible to present random-dot stereograms via the VISIDEP method, led to an entirely new concept. Suppose that two stereoscopic images, either left and right or up and down, were broken into many separate parts and then recombined to form two new images, each containing information from both of the original stereoscopic views. Then these new images, which are just linear combinations of the originals, could be thought of as a pair of random-dot stereograms. When viewed using the method of alternating pairs, the depth should still be visible. Furthermore, since each of the new images contains information from both of the original views, objects that were located beyond the convergence point would appear doubled, as in normal vision, and the distracting motion seen in the simple alternation technique should disappear. The use of composite images should restore stability to the image and still allow the sensation of depth.

A simple realization of the composite image can be obtained in standard NTSC television by switching between fields [JONE85]. One of the composite images consists of a full-frame image in which the first field derives from camera or source A and the second field derives from camera or source B. The other composite image is composed of a first field from B and a second field from A. These two images may then be alternated at a rate of ten changes per second to produce the desired depth effect. With NTSC, the alternations follow every third complete frame. Because the interlaced fields are displaced vertically on the CRT, horizontal parallax between the cameras is preferred in order to eliminate unnecessary vertical movement in the image.

The image that results from the composite technique is remarkably stable, as expected. Depth is preserved in the display, but it is somewhat suppressed. Thus, a greater separation is called for between the cameras. Further smoothing is achieved by using the composite technique and the mixing technique together.

A difficulty with this simple embodiment of the composite technique arises because of the manner in which the television image is written to the screen in NTSC. At the times in the sequence of fields when the switch is made from one source to the other, successive fields can come from the same camera, momentarily creating a single frame from that source. This effect contributes an undesirable kinetic element to the picture. However, if the composite images were presented serially, as in motion pictures, where there is no screen persistence and no interlaced writing of the image, this effect would never be present, and the image would be much improved.

12.5 Variations Using Computer Graphics

When computer-generated images are used to provide the source material for VISIDEP images, the problems of brightness, color, and size associated with the two-camera video system can be entirely eliminated. Simulations of the original AP method have been achieved by Barham, Harrison, and McAllister [BARH90] using a perspective projection with one eyepoint, and then rotating the entire scene about a horizontal axis that is perpendicular to the line of sight and contains the point of convergence of the two video cameras. This rotation provides the vertically aligned parallax information to the viewer.

In their work, Barham, Harrison, and McAllister sought to improve on the AP technology using computer-generated images of scenes with no overlapping objects, no color cues, no texture cues, and no shading cues. They then attempted to reduce the distraction of the rocking motion through three techniques. The first technique was to employ a complex texture pattern at the rear of the scene to provide a point of reference for all other objects. The addition of this reference plane greatly improved the perception of depth. The second technique consisted of rotating each object about its own axis instead of using a single rotation axis. That attempt proved disappointing, as the individual rotations did not convey the same information as a parallax shift associated with rotations about a common axis, and thus did not help the viewer to judge interobject distances.

In our work with video input we produced acceptable images with the cameras converged from near the center of the view volume along the optic axis to the rear of the view volume. In their computer-generated scenes, Barham and colleagues [BARH90] found that the least distracting scenes were those with the rotation axis located within the deepest third of the view volume. They found that placing the rotation axis near the front of the scene did not reduce the depth enhancement, but it did increase the motional distraction. Mayhew [MAYH87] has reported that allowing the convergence (i.e., the rotation axis) to follow the nearest object in the scene improves the stability of the image. His observations were made with motion picture animation derived from computed images consisting of a relatively simple system of moving objects with no reference plane. More recently, Mayhew [MAYH90] asserted that the technique of front convergence improves video images as well. We have been unable to confirm this claim, but our own experience and that of Barham and colleagues contradicts his allegation.

The third technique investigated by Barham, Harrison, and McAllister [BARH90] was to simulate the use of three or four cameras all slightly displaced to provide three or four viewpoints. The effect of more cameras was to smooth the transition between images and reduce the irritation of the rocking motion. The four-viewpoint sequence was judged better than the three-viewpoint sequence or the two-viewpoint sequence. The four-image presentation was played in a ramplike sequence of frames in the order 1,2,3,4,1,2,3,4,1,2,3,4,1, and in a smoother sequence in the order 1,2,3,4,3,2,1,2,3,4,3,2,1,2. As expected, the second ordering produced better results.

12.6 Applications

Television Systems

The immediate applications of AP imaging are in television. The technique allows the origination of live, real-time signals for transmission or recording in full color with no signal degradation. It can be used for broadcasting, for closed-circuit use in training or research, or for recording and later playback. However, we have long felt that the primary applications of AP lie beyond the realm of broadcast television and entertainment, and we have considered its application in other areas.

CAD/CAM and Robotics

A second important area of application is computer-generated imaging. AP is one of two viable methods for achieving 3D images on computer displays ([HODG85], [BARH90]). The fact that there is no requirement of special glasses or monitors makes this system simple to employ and more user-friendly than other systems currently available.

The value of three-dimensional imaging in CAD/CAM operations has been recognized for some time [TEIC85]. VISIDEP techniques offer the potential for improved visualization of 3D images in CAD/CAM. Digital control of the angular disparity between the two views provides direct control of both depth and motion. When this control is combined with the mixed technique, precise mastery over the image is available to the operator.

There is also a great potential for enhancement of medical imaging. Preliminary work with x-ray pictures has shown that these images can be rendered in depth by the VISIDEP process. However, because of the soft edges and general lack of texture, x-ray images are not as easy to interpret as television images of visual scenes. Applications to MRI and computerized tomography are yet to be studied, but there is no fundamental reason why they should not also prove successful.

Potential applications to robotics have been investigated [MCLA85]. An industrial robot could be designed to provide quantitative measurement of depth and simultaneously give a visual illusion for a human operator, permitting three-dimensional remote viewing of hostile environments.

Remote Sensing

Another variation of the VISIDEP technology is specially designed for remote-sensing applications [MCLA89]. This technique uses a single moving source (camera or other sensor) to produce a succession of frames taken from different locations. Comparison of one of these frames with another frame nearby reveals considerable overlap in their images; they are stereoscopically correlated. Stereoscopic viewing of these frames produces a three-dimensional image.

A 3D image also results if these two frames are presented sequentially to a viewer using the alternating-pair technique. When the two images are carefully registered, the resulting image has depth. It is possible, therefore, to take the individual frames made with the moving camera, reorder them, and produce a new sequence of frames that will be perceived as three-dimensional.

Assuming constant-level travel of the source at a selected height, frames can be chosen to give a desired amount of depth based on their separation in the sequence. This separation corresponds to the time interval between the frames, and its choice determines the angle of binocular separation between the images. Note that in comparison with the two-camera system described earlier, this system uses a temporal displacement of the images.

Notice also that we are not merely using monocular movement parallax. We are incorporating something more, an effect that can produce an image with depth even when there is no discernible depth from monocular movement parallax alone. This distinction should not be forgotten.

Initial tests of this single-camera system have been carried out photographically [MCLA87] because of the ease of manipulation and the lack of a real-time requirement for feasibility testing. Our tests show a definite enhancement in image depth. An advanced electrographic system combining video and computer technologies can be designed to provide near-real-time visual inspection and analysis of remote locations. For example, an aircraft-mounted camera flown at constant speed and altitude produces a steady stream of images that can be programmed into a three-dimensional view. The time delay in the signals compared for the depth enhancement is a function of the speed, altitude, and angle of binocular disparity required for the creation of the 3D image.

Figure 12.3 provides a basic conceptual illustration of a single-source moving-platform system. A primary advantage of this single-camera system is the obvious reduction in size and weight, along with the removal of problems in color matching and light loss that occur in a two-camera system. A single source (camera) provides consistency of color, light, and geometric shape, which is difficult to achieve in a two-camera system. The illustration shows an aerial platform, but the aircraft may be replaced with any constantly moving platform operated in a land, air, or marine environment.

The signal from the single source is split into two identical signals, A and B. After splitting the signal, the operator introduces a delay in signal B

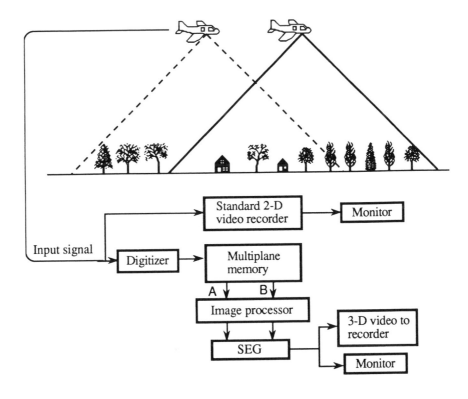

Figure 12.3
Single-camera moving-platform system.

based on the angular disparity required for the desired depth effect. This time delay can be expressed in terms of video fields rounded off to the nearest integer. (For NTSC there are sixty fields per second.) The temporal delay in signal B relative to signal A provides the requisite stereoscopic information because the delayed signal corresponds to a different spatial location of the camera.

The temporal delay can be achieved in several ways. One method is to digitize the images and store them as individual frames in a multipage memory. Successive update frames are stored sequentially in a wraparound fashion. Images are read out to a dynamic display in a sequence similar to that used for the film test just described. However, the electronic processor allows for direct viewer input to control the amount of offset by adjusting the frame separation to suit the needs of the task at hand.

In the single-source system, the critical alignment of the images is done electronically. Standard image-processing techniques can be used to align the images and to correct for image distortions due to changes in angle, for example. A computer could handle these adjustments rapidly, making near-real-time changes in the system. By alternating the overlapping sections of the images in sequence A and B, the system creates a new image, C, which contains the three-dimensional information. The result can be viewed directly, recorded, and analyzed just like any other signal.

Thus, AP technology is readily adaptable to remote sensing. Both the dual-camera system and the single-camera system can be used to provide 3D data in settings where traditional 2D systems are in use. Almost any situation in which the information is presented visually is a candidate for depth enhancement. Size and scale are no longer major considerations, as the two following examples demonstrate.

Consider the potential of fiber optics in the generation of the initial signal. Images obtained through coherent fiber bundles could be used to measure depth and simultaneously enable visual inspection of areas far too small to be entered by humans or roaming robots. The interiors of reactor tubes, pipes in chemical plants, and similar areas could be seen with such a system, providing data in both visual and recorded forms. Special adaptations could be made for medical purposes.

Another system could be devised to provide data from operations of a much larger nature. The evaluation of damage to undersea structures is important to shipping, oil drilling, and other industries. Similar systems could be used to gather geological data both above and below the surface of the earth. Camera systems can be designed to work in hostile environments where humans cannot survive, and they can then send a near-real-time 3D image to a sheltered environment anywhere in the world. The cameras could, of course, be remotely controlled from the safe location, including control of the parallax.

The single-source system has its most obvious applications in aerial reconnaissance, which is illustrated in Figure 12.3. Uses include aerial data (image) collection in low-altitude viewing, that is, from normal aircraft, such as might be used in military surveillance, tactical reconnaissance, and forward observation through the use of drones. High-altitude viewing from satellites would allow the observation of gross features. It would be especially useful in meteorology, as weather systems and cloud formations could be imaged in both visible light and infrared.

Both of the methods described here can be used with image information originating from different sources, including infrared and x-ray sources, as well as direct computer-generated images. Thus, the applications extend over a wide range. Additional applications include quantification of digitized image information to extract measurements of the object being imaged [MCLA85].

Medical Endoscopy

Medical endoscopes have become indispensable tools for diagnostics and minimally invasive surgery. One feature common to all of these is the presentation of a monocular view to the eye of the clinician. In contemporary practice, a miniature video camera head is attached directly to the end of the scope, and the resulting image is displayed on a color television monitor. The available cameras are capable of providing bright, high-quality, full-color images. Because of the perspective distortions due to the short-focal-length lenses required, the location of an object in depth within these monocular images is often difficult to determine.

Through the use of the AP alternating-frame technique, endoscopes for surgical and other applications can be built that present real-time three-dimensional images on a video screen. Because they are autostereoscopic, these images can be viewed directly without glasses or other aids [JONE90].

At the request of an orthopedic surgeon, we made some tests to determine if a scope could be built within the constraints of current practice in arthroscopy. After these investigations and after discussions with a scope manufacturer, we designed a laparoscope with dual imaging systems [MCLA90]. A scope for three-dimensional viewing was then constructed to the same size and scale as laparoscopes currently in use by physicians. The result was a laparoscope 11 mm in outside diameter with objective lenses directed along the longitudinal axis of the scope tube. Ample space was available for fiber-optic conduits for bringing light from a source to illuminate the region being examined. The physical layout of the fiber bundles and the objective lenses of the prototype instrument is shown in Figure 12.4. The center-to-center separation between the two lenses was 6 mm, and they were mounted so that their optic axes converged at a distance of 60 mm beyond the end of the scope. At that distance the centers of their fields of view coincided. The angular separation between the two views was

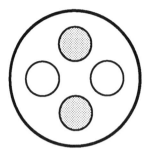

Figure 12.4
End view of the laparoscope probe, showing the positions of the fiber bundles (shaded circles), which transmit light to the subject, and the objective lenses (open circles) of the imaging system. The outside diameter is 11 mm.

nearly 6° at the point of convergence. Such angular disparity is well within the requirements for good depth sensation [JONE84].

The two images were combined so that they could be directed along a common optical path and then be brought out through a standard eyepiece, allowing the emergent image to be seen directly with the eye. A system of prisms and a beam splitter were used to provide folded optical paths combining the two beams in a manner similar to that used in binocular microscopes to split a single image so that it can be sent to separate eyes [JENT14]. Electro-optic shutters were placed in the two individual beam paths where they could be alternately opened and closed so that only one beam at a time passed through (Figure 12.5). The scope was then connected to a video camera by means of a standard adapter.

Two possibilities remain for presentation of the three-dimensional images. The shutters may be switched at intervals of three frames (six fields) to generate an autostereoscopic AP image on the monitor screen. In this mode there is no need for shuttered glasses, and the requirement for horizontal parallax is eliminated. Alternatively, the shutters may be synchronized with the camera to switch on alternate fields. Then the video monitor is viewed through shuttered glasses so that alternate fields are presented to separate eyes. The result is a traditional binocular stereoscopic image [STAR90]. However, in this mode, care must be taken to ensure that the resulting image parallax is strictly horizontal; otherwise the eyes will be unable to fuse the images to create the 3D view. Once aligned, the scope and camera

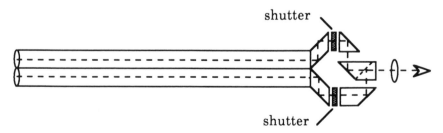

shutter

shutter

Figure 12.5 Diagram of the prismatic system used to combine the two beams along a common axis. The position of the shutters is indicated.

must not be rotated relative to each other; rotations must be accomplished so that camera and probe move as a single unit. This arrangement maintains the horizontal parallax on the screen, but the image orientation that is so desirable in some applications is lost.

One distinct advantage of presenting the image using the AP technique is that the user is free to rotate the scope to any angle about the symmetry axis relative to the orientation of the camera head. This is especially important in other endoscopic applications, such as arthroscopy, where the center of the field of view usually is directed at an angle 30° from the longitudinal axis of the scope tube, and the lens system affords a wide-angle view of approximately 120°. Such a system permits the user to look around in different directions by simply rotating the scope. If the scope is rotated while the camera head remains fixed, the 3D image generated maintains its orientation on the screen even though the scope is rotated about its axis.

The endoscope described here is capable of providing 3D images with commercially available television equipment. When operated in the AP mode, it may be used interchangeably with cameras and monitors already in use to present an image requiring no other viewing aids. When it is operated in the field-sequential mode, additional equipment is required, either in the form of shuttered glasses or in the form of passive glasses and an active polarizer that covers the screen [BUZA85].

12.7 Frequently Asked Questions

Whenever the VISIDEP technology is presented to an audience, the same questions are often asked. Because of their frequency, we wish to respond to them here as well.

Because there is sometimes a discernible motion, or wobble, in the image, people often ask, "What causes the wobble, and can you get rid of it?" The origin of this motion lies in the geometry of the optical system required for creating two binocularly related views. (Even if the disparity is vertical, we still refer to it as a binocular disparity.) There will be some portion of the object scene at which the optic axes of the cameras converge. In other portions, the optic axes do not converge, and the resulting images are slightly disparate. When the two views are alternated, the disparate parts of the image are perceived to move. Efforts to reduce this unwanted image motion have been successful. The use of carefully constructed optics and broadcast-quality cameras have eliminated all of the other unwanted distractors from the alternating images, resulting in a pleasant reproduction of three-dimensional space. Nevertheless, it may be impossible to completely eliminate the motion in alternating-pair presentations without also eliminating the depth sensation. However, all 3D systems have some drawbacks, such as the need for special glasses or other particular viewing equipment, the repetition of images in lenticular displays, and so forth. For many applications, parallax motion that cannot be eliminated from the AP images is offset by the system's simplicity and avoidance of special viewing glasses or devices.

Another frequently asked question concerns monocular motion parallax. People knowledgeable of 3D imaging often ask if the effect seen in AP is simply motion parallax, particularly if they have seen imaging in which the camera was in motion during the scene. However, one has only to look at the research describing the sequential presentation of disparate images to realize that there is more involved than motion parallax. Comparison of results from a single moving source with the results from the alternating-pair method show a substantial difference in the amount of perceived depth [MCLA87]. Several striking features should be noted. With AP there is a depth illusion even when the camera system is stationary. However, if the camera is in motion, if there is motion in the scene, or even if the camera lens is zoomed, the perception of image motion due to alternating viewpoints is suppressed, probably because of the overall complexity of the viewing situation. In any case, the depth illusion is maintained.

A third question concerns the use of vertical disparity. Our research has shown that not only is vertical parallax permissible; it is preferred. For the same effective separation between cameras, vertical separation gives greater stability to the image, and in the case of complex, unfamiliar images, vertical parallax reduces errors of misinterpretation. The use of vertical parallax is permissible because rather than providing separate information to the separate eyes simultaneously, we provide the necessary information to the brain via both eyes, and since both eyes see the same image, it doesn't matter which way the disparity is oriented.

12.8 Summary

The alternating-pair method of presenting three-dimensional images on flat screens has been tested and used for a number of years since its introduction in 1982. New applications and methods have been developed by the authors and others for television, motion picture, and computer-generated source material.

The AP approach has specific advantages over other existing systems:

1. The three-dimensional images are viewable without aids to the eye, and viewing position is unrestricted. Tilting or lateral motion of the head does not affect the depth perceived, unlike the case with most stereoscopic systems.

2. The images can be produced, recorded, and retrieved with standard, readily available equipment. Presentation of the images does not require specialized monitors, screens, or other hardware.

3. The technique is adaptable to all imaging systems, including computer-generated imaging.

4. No color loss or light loss occurs in the playback system, as it does in systems that employ polarized optics or color filters. AP works equally well in both monochromatic and full-color presentations.

5. AP works with both static and kinetic images. It does not require continuous rotation or translation of the image to create the depth illusion.

There are also some disadvantages to the AP method of presenting 3D visual information:

1. There is a slight rocking motion of the images that can be distracting, especially if a static scene is being shown.

2. There is no "lookaround." That is, motion of the observer's head produces no additional parallax and reveals no new information. However, this is true of most stereoscopic systems.

3. When two cameras are used for generating the images, there is a critical requirement that their images be exactly matched in size and shape, brightness, color balance, contrast, and resolution. This disadvantage may be eliminated with the use of single-camera systems or computer-generated source material.

4. Observers cannot roam through the depth of three-dimensional images by altering the convergence of their eyes.

The potential of the alternating-pair method of three-dimensional imaging has not been fully realized. The realization of this potential will come about when the system is in the hands of the end users.

13

Volumetric Three-Dimensional Display Technology

R. Don Williams

13.1 Introduction

This chapter provides an overview of the technology available to create a physical-volume 3D display, one presenting images having height, depth, and width. These images occupy physical space and are not a 2D rendering of a 3D image.

Perspective CRT displays provide limited information for spatial tasks. In a series of reports, a research team at NASA Ames Research Center characterized the errors made when observers reconstructed 3D spatial layouts from 2D perspective projections ([MCGR84], [MCGR85], [GRUN86], [GRUN88]).

Volume information can be displayed, in limited form, on two-dimensional flat displays with perspective cues, stereoscopically as detailed in other chapters of this book, or as a physical 3D volume as described in this chapter. Although the parallax cues from stereoscopic presentations suffice for many tasks, some individuals have various degrees of stereoblindness. In addition, multiple-person (team) tasks can require many simultaneous views or the "lookaround" capability, in which case a true volume display is preferred. A volume display with parallax and motion parallax cues should also greatly reduce estimation errors and improve task performance for many applications. This chapter reviews the various technologies for implementing volume displays and some of the perception issues and standards involved in such displays.

13.2 History of Volumetric 3D Technology

A number of attempts have been made to provide volume displays with true 3D images that closely resemble real-world objects. The history of the technology for volume displays includes several unique approaches. Volume displays can be divided into two basic categories. *Solid volumetric* displays use physical elements at each specific (x, y, z) location as an emissive luminary. *Multiplanar* displays depend on a moving element or multiple elements that occupy each of the (x, y, z) positions in the volume over time. The receptors in the eye have a temporal persistence that fuses the light emitted from the moving (x, y) element to create the volume image.

Solid Volumetric Displays

The goal of a solid volumetric display is to provide emissive voxels, or three-dimensional pixels, at a large number of (x, y, z) locations. Assuming that practical applications require a minimum volume of $64 \times 64 \times 64$, 262,144 voxels need to be spaced so that the eye cannot detect a separation between two emitting voxels, a requirement of about 2° of angular subtense. This corresponds to a voxel every 1.15 mm, or a volume size 74 mm on a side. These requirements are difficult to implement with today's technology. Several interesting attempts have been made using gelatins, gases, and glass spheres.

The story is told of one researcher who filled a large aquarium with transparent gelatin material. Using a mechanically positioned stylus, the researcher deposited ink traces in the volume. This display needed no refresh, as the gelatin held the ink in place. A major problem obviously arises when a new image is to be traced out in the gelatin and the old display must be flushed from the display tank. However, such a simple volume plotter would be effective for static data to show complex interactions in the volume. This example helps to illustrate the concept of the solid volumetric display.

Another, more serious approach used a gaseous volume enclosed in a sealed glass container. The gas could be excited by the intersection of two orthogonally scanned laser beams. In this work, funded by the Office of Naval Research and conducted by R. H. Barnes at Battelle Columbus Laboratories, the researchers developed a method to generate a fluorescent spot at the intersection of two laser beams within a volume of suitable material by the sequential excitation of fluorescence. Over a six-year period, Battelle evaluated the display system concept, materials, and applications for the 3D display. The group determined that the most promising gas was a diatomic vapor, iodine monochloride (ICl). They concluded that, with 1975 laser technology, they could build a 150-spot display and that the power levels for a 1,000-spot display were available, but that increased laser pulse repetition rates with high peak-pulse intensities were needed.

This work resulted in U.S. Patent 3,829,838, issued to Eric J. Korevaar and Bret Spivey, and is further detailed in [LEWI71] and [BARN74]. Figure 13.1 illustrates their approach, which uses two orthogonal lasers scanned to intersect at (x, y, z) points in the volume. A first laser source excites a small volume of gas to the first excited state, and the second laser excites the gas to a second excited state, which relaxes to emit visible light from that (x, y, z) point, creating an emissive voxel. They suggest that the excitation medium could be composed of molecules rather than atoms, and that the medium could be a transparent liquid or even a solid rather than a gas.

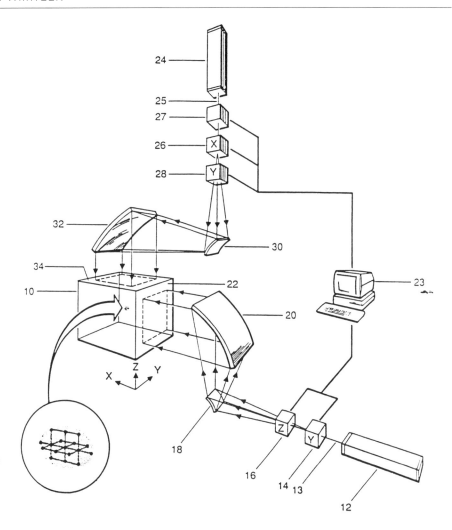

Figure 13.1
Gaseous volume with two orthogonal lasers.

Another interesting attempt to provide a physical-volume display is described in U.S. Patent 4,023,158, entitled "Real Three-Dimensional Visual Display Arrangement," by Donald G. Corcoran and assigned to International Telephone and Telegraph Corporation. This system is based on the levitation of tiny glass spheres using laser light. As shown in Figure 13.2, the beams are modulated and scanned in x and y, and the z dimension, which is the height of a glass sphere, is determined by the power of the laser. We don't know if this system was ever prototyped or further developed. This system attempted to meet the goal of a solid volumetric display by levitating physical objects in a volume of space and having them emit light. Like the gaseous approach, this system puts the burden of the large number of points on the scanner technology.

Since little effort has been devoted to these approaches in the last decade, they seem to have been unworkable. Perhaps with present-day laser and

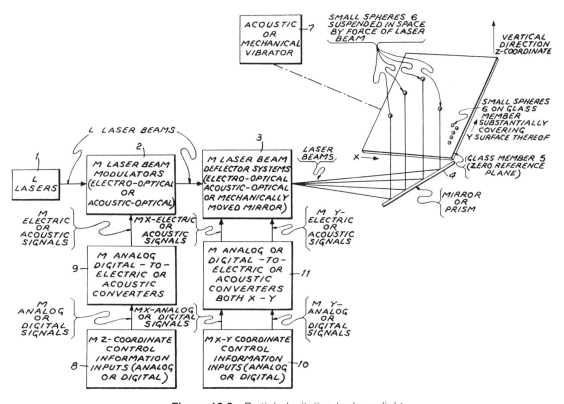

Figure 13.2 Particle levitation by laser light.

scanner technology, these approaches can eventually prove technically viable. Until these solid-volume displays become technical reality, the most feasible near-term solutions appear to be multiplanar 3D displays.

Multiplanar Volume Displays

Multiplanar displays depend on a moving element that occupies a single plane when not rotating. To describe a volume (usually cylindrical in shape), the plane is moved so that it occupies each of the (x, y, z) positions in the volume over time. The rods and cones in the retina of the human eye have a temporal persistence because of a processing delay, and this causes the eye to fuse the light emitted from the moving (x, y) element into a volume image. The plane in a multiplanar display can take various shapes and can emit light from some internal source or act as a projection surface. The following review of multiplanar designs illustrates the diversity in these systems. However, they all have the goal of projecting points of light from specific (x, y, z) locations using temporal movement.

Probably the most widely known volume display is the varifocal mirror system called SpaceGraph, described in Chapter 11. For another varifocal design, Thomas O'Brien was issued U.S. Patent 4,639,081, which describes a gimbaled flexible membrane mirror. This 3D display system was assigned to Toshiba in 1987. An earlier patent issued in 1968 to Chris S. Anderson used a pair of articulating mirrors to reflect a CRT image at various focal distances. All three of these varifocal multiplanar displays project a sequence of 2D frames while using the varifocal mirror to change the image focal distance. When the frames are synchronized with the focal planes, the viewer sees the frames stacked in volume image. Varifocal mirrors—whether vibrating, flexible surfaces or moving plate mirrors—create image distortions that must be corrected in hardware or software.

The multiplanar approach patented by Jack Fajans (U.S. Patent 4,315,281) is known as the *xyz* scope. It uses a positive lens rotating at 600 rpm to create an oscillating point image along a line segment. The light pulses radiate from image points on the segment when the image source is flashed at corresponding lens angles. A 3D matrix of voxels is generated by pulsing the image source, a CRT or LEDs. This device is reported to generate real-time images (10^5 pixels/second) in space in normal room lighting without the need for special glasses. This technology was first shown by Fajans in 1979. A later version of this system appeared under the product name Triology and is covered by U.S. Patent 4,692,878, issued to Bernard M. Clongoli, president of Ampower Technologies. The rotating lens is said to cause aberrations as the viewer moves out of the center axis of the lens.

Figure 13.3 shows a volume display system developed by Edwin P. Berlin, Jr., and assigned to the Massachusetts Institute of Technology (U.S. Patent 4,160,973). This system uses a 2D array of LEDs that are rotated to scan a 3D volume. The resolution of this volume display is a function of the number and density of LEDs mounted on the rotating planar array, the speed of rotation, and the rate at which the LEDs can be pulsed.

An alternative to rotating the projection lens or using an emissive planar array is to rotate a 2D surface that receives a projected beam to provide emissive voxels from any point in a volume. These multiplanar displays create an (*x, y, z*)-addressable display volume by rotating a light-receiving planar projection surface. These rotating 2D surfaces describe a 3D volume (cylinder) that is used to display a volume image. The rotating multiplanar 3D laser display generates 3D images in real time that are viewable from all angles and thus can be viewed simultaneously by many users.

In 1964 a patent was issued to R. D. Ketchpel (U.S. Patent 3,140,415) providing for a rotatable screen that is mounted in a vacuum and has the ability to luminesce when bombarded with electrons (Figure 13.4). A standard CRT deflection yoke allows addressing of the rotating phosphor disk. The view of this rotating multiplanar surface depends on the clarity of the glass enclosure and the translucency of the rotating screen. Another image quality issue is the interaction between the phosphor decay rate and the speed of the rotating screen.

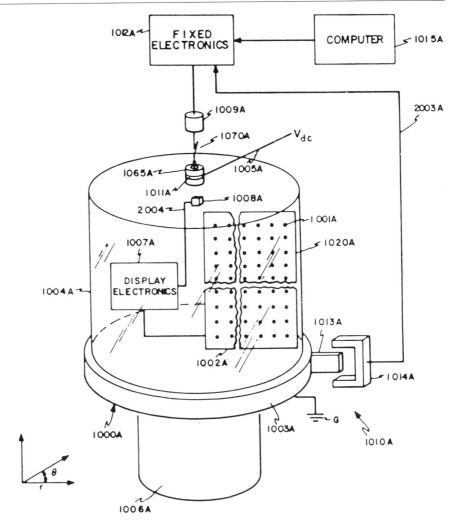

Figure 13.3
2D array of LEDs
rotated to scan a
3D volume.

A 1976 patent issued to Wendell D. Chase and assigned to NASA uses a series of successively thicker pie-shaped glass sections (Figure 13.5). Frames of a strobed image are projected on the back of the glass slices, and the image is viewed from the opposite side as the glass sections rotate. The effect is a 3D image for a single viewer with a limited range of view. The variable thickness of the glass sections can have a dramatic effect on the luminance uniformity across 3D frames. Also, the definition of the edges of the sections can cause distortions.

A further adaptation of a rotating projection surface was devised by Yamada [YAMA86]. To simplify the processing of video signals, a rotating projection screen was designed to move in the shape of a sawtooth over time. Figure 13.6 illustrates the Archimedes' spiral projection surface used to produce this sawtooth effect. Two segments of the screen are fixed along

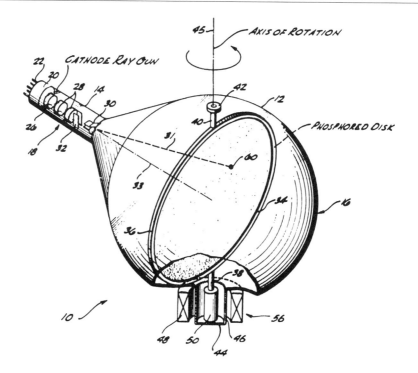

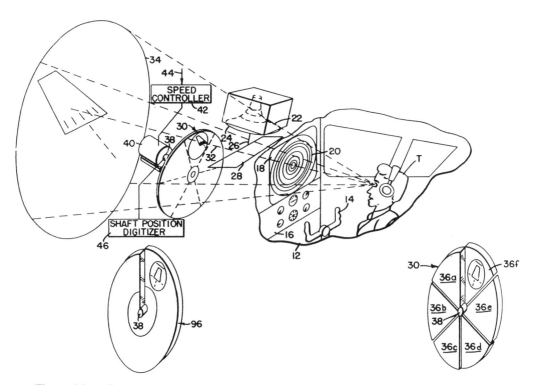

Figure 13.5 Frames of strobed images projected on the back of pie-shaped glass slices.

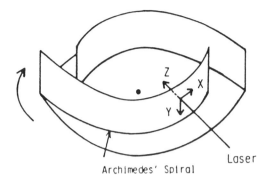

Figure 13.6
Archimedes'
spiral projection
surface.

each half turn of the Archimedes' spiral drawn on the disk. A raster-scanned laser beam projects sequential cross-sectional images onto the Archimedes' spiral from the side of the rotating disk. Design of the spirals in this device seems to limit the viewing angle. The discontinuities of the spiral edges could also introduce image effects.

Another rotating-projection-surface approach, suggested by Rudiger Hartwig of the University of Stuttgart, uses a series of static, instantaneous images projected onto a single-turn helical surface. This work reports a true volumetric vector image based on a laser beam parallel to the axis of the rotating helix (Figure 13.7). Random wirelike images are seen inside a glass cylinder [BRIN83].

U.S. Patent 4,922,336, recently issued to Roger Morton and assigned to Eastman Kodak, describes a rotating helical projection surface much like the Hartwig design. The unique aspect of this patent is the use of an anamorphic lens to correct for the variations in the focal distance of the

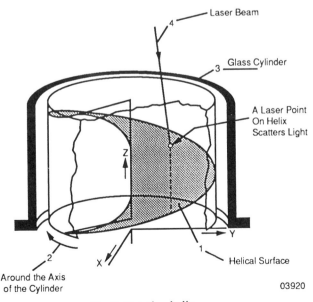

Figure 13.7
Rotating-helix
projection
surface.

Fig. 8. Rotating helix.

various points on the helical surface (Figure 13.8). This special correction lens rotates with the projection surface to form a projected CRT 2D image.

Over the last several years, Texas Instruments has developed an autostereoscopic multiplanar volume display. Figure 13.9 shows an overview of this rotating projection multiplanar display technology. The display surface, a double-helix transparent display disk, rotates to fill the display cylinder. The surface rotates at 600 rpm, creating a cylindrical volume where the 2D images are fused by the viewer's eyes. The entire volume of the cylinder is displayable except for a small cylinder in the center of the volume. The display system uses a laser beam modulated up to 10,000 Hz and synchronized with the z-dimension displacement of the rotating disk. The beam is scanned in x and y, and synchronization of the modulation with the rotating disk defines the z dimension. The disk is translucent, providing persistent 2D slices that are fused by the viewer's eyes to create volume 3D images.

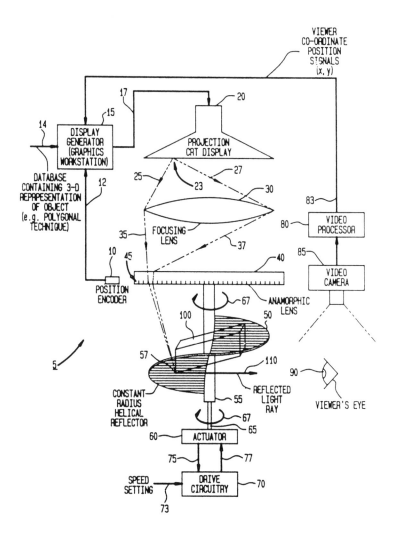

Figure 13.8

Rotating helix with anamorphic lens.

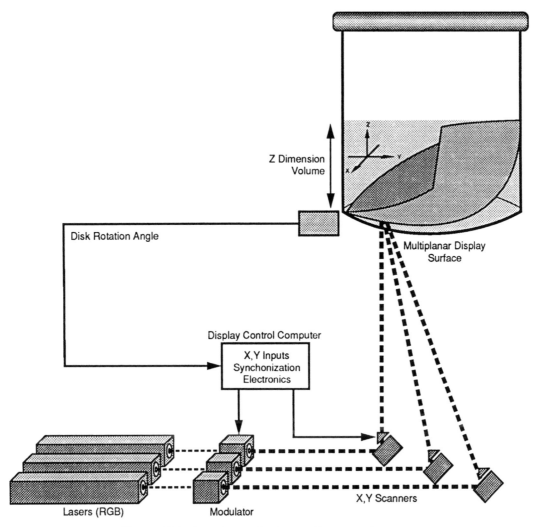

Figure 13.9 Rotating projection multiplanar volume display.

The computer detects rotation of the disk from an optical sensor on the shaft of the motor. A volume display 36 inches in diameter and 18 inches high using three colors and approximately 10,000 displayable voxels was delivered in January 1992 (U.S. Patent 5,042,909). This technology has been marketed under the name OmniView and is being offered as a custom system (Colorplate 13.1).

Rotating multiplanar technology shows promise of having some advantages over existing 2D or 3D displays for many volume visualization tasks. These advantages include the creation of large volumes (a 36-inch-diameter disk has been prototyped and can be theoretically extended to a very large volume); the ability to provide all-angle views, or the lookaround property (internal views can be produced by clipping); and flexible design (the size and effectiveness of the display workstation environment can be optimized for a particular task). For example, a battle management task would require a large, flat cylinder to model the battlefield, and a finite element analysis might require a tall cylinder to display the stress points on a long-wing structural model. Each of these tasks could be accommodated by the flexible nature of this technology.

In addition, such displays can have dynamic resolution; the display resolution can be dynamically modified by redirection of the scanning system outputs, and these can be interlaced with another scanner's output to increase the resolution. The voxel density can be increased by adding more scanning units, as the display resolution is not fixed by the display surface. Finally, as previously indicated, the displayed image has parallax and motion parallax properties similar to real-world 3D objects. The primary disadvantages of such displays include the fixed transparency of the image and the difficulty of producing high resolution.

13.3 Human Visual Issues for Volume Display Technology

An extensive body of literature has been developed concerning image quality issues for 2D displays, particularly CRTs. The American National Standard for Human Factors Engineering of Visual Display Terminal Workstations (ANSI-HFS-100) documents the minimum visual image quality acceptable for office tasks and defines image quality measurement techniques. These requirements provide a baseline for evaluating the visual performance of the new volume display technology, but they must be extended and revised to consider the third dimension of the display. The visual issues associated with the latest multiplanar display technology can be separated into two categories: issues resulting from the use of a coherent light source (e.g., a laser) and issues relating to the (x, y, z) dimensionality (volume) nature of the voxels.

Laser-Based Luminance Issues

The laser is an excellent high-power luminance source for display devices. With recent improvements in the power/cost ratio, lasers will enjoy rapidly increasing use as display luminance sources. Also, the availability of visible laser diodes will greatly reduce the packaging, power, and cooling requirements. The laser has long been used in the entertainment field because of its high luminance, contrast, and image persistence. However, the use of lasers presents several unique image quality issues for viewers. Some of the most important issues are reviewed here to provide guidelines for future research.

Excessive Contrast. Too much contrast causes the displayed image to become a source of glare, which has contributed to subjective reports of visual fatigue. Standards recommend display modulation in the range from 0.5 to 0.95. Research on CRT technology has shown that if the contrast is too high, two effects are observed. First, the image appears to float, with no visual background to anchor it. Second, the spot appears to bloom, which causes the image to be less sharp. The requirements for image surround ratios in 2D displays need to be validated for volume display technology.

Laser Speckle. A unique characteristic of coherent laser light is the speckle patterns created when the beam is reflected from a projection surface. Laser speckle can occur when the projection surface irregularities in depth are greater than one-quarter of the wavelength of the incident light. Several methods, such as screen movement, exist for reducing laser speckle in 2D displays. The existing literature for laser displays needs to be extended to address the use of the laser in multiplanar displays.

Color Perception. Perhaps because the laser produces a very narrow band and highly saturated color, two problems may occur. Accommodation in the visual system varies as a function of wavelength or hue. For narrow-band filtered light, the differences in focus are about 2.5 diopters. The eye's minimum depth of field is about ±0.3 diopter. The use of high-luminance or high-contrast images generated by a laser could have an effect on the accommodative response of the eye, especially if the cues are not sufficient for the eye to focus.

The second issue is color distortion. The chromatic aberration of the eye could be exaggerated by highly saturated narrow-wavelength light. Thus, red and blue objects at the same depth can appear to be at different depths in the display. The interaction of these chromatic effects with volume parallax and motion parallax cues is an important area for further study.

Beam Shape. The image quality of a laser-based display is a function of the beam shape. The beam shape is determined by the beam width and the intensity distribution across the beam diameter. Cylindrical and annular beams have a better modulation transfer function than the Gaussian spot of a CRT. Beam shape involves key trade-offs in the reproduction of lower versus higher spatial frequencies.

Volume Display Issues

Many of the image quality issues can be addressed using a direct extension of the 2D display literature and the existing image quality metrics. Others require new psychophysical models and verification methods. Because of limitations in scope, this discussion focuses on selected critical issues in volume displays.

Luminance. The required minimum luminance is a function of the background luminance, the contrast enhancement filters used over the display volume, and the interpositional voxel density and transparency of the image. A minimum luminance of 35 cd/m^2 is recommended by ANSI HFS-100. The maximum luminance for eye safety has been documented in FDA21CFR Ch.1 part 1040.10 and standardized in ANSI Z136.1-1986. These documents, however, do not consider long-term viewing of images from coherent light sources (lasers).

Luminance Uniformity. The voxel luminance can vary over the display volume. The ANSI standard requires less than 20 percent contrast between adjacent pixels when there are fewer than thirty per degree of angular subtense. Also, the display luminance should not vary more than 50 percent center to edge. A metric for luminance uniformity is needed to define requirements for the z-dimensional variations and to identify any interactions with absolute luminance level, contrast, or color.

Color Registration. Color can be achieved in a volume display by a number of design approaches. Acousto-optic scanners used for high-bandwidth projection into the volume display must be designed for a specific wavelength of light. This wavelength specificity and the need to use multiple scanners for bandwidth multiplication require that the various colors be combined optically after the 2D image is created in the scanner. The optics must exactly position the scanned image on top of the image from other scanners for a full-color volume display. The accuracy of this convergence determines color purity. Additionally, the timing of the scanned image of each voxel is critical to achieve z-axis registration of the color primaries. By time-division-multiplexing each voxel in the three primary colors, a full color palette can be obtained. Alternatively, with scanners that are wavelength-independent, the primaries can be combined before the scanners using fiber-optic combiners, eliminating the color registration problem.

Color Contrast. The ANSI standard defines a color contrast of 40 ΔE (CIE L*u*v* scale) as the minimum separation needed for color discrimination. Volume displays should require a simple extension of this measure with an added ΔE$_s$ to overcome the effects of translucence on color. More systematic research is needed to define the influence of translucence on color discrimination.

Voxel Stability Issues

Voxel Definition. Two-dimensional display research has determined that a square pixel yields the best display image quality. A simple extension of this literature to volume displays suggests that a cubic voxel would provide the best volume image quality. However, the current X-shaped voxel is typically perceived as spherical. Also, the perceived voxel volume is dependent on the viewing angle. Voxel definition is an important element in volume image perception. Voxel definition may interact with the volume object type and definition. Lines may need one type of voxel and complex icons another.

Image Translucence. A unique characteristic of multiplanar volume displays is the ability to see through the image to view the points behind it. For some applications, this feature enhances visualization of complex data. For other applications, this is an unacceptable limitation in the technology. The ability to vary the transparency of the imagery has numerous implications for medicine, some artistic displays, and visualization applications.

(x, y, z) Point Stability (Jitter). The movement of the voxel in (x, y, z) space is an important element in overall image quality perception. This 3D jitter may affect the viewer's ability to accommodate on the image. Image jitter is perceived as flicker and compounds any negative temporal perceptions of a volume display system. The ANSI standard requires a jitter of less than 0.0002 mm per millimeter of viewing distance. It is unknown if this requirement holds for the z dimension. A metric for 3D measurement of this movement over time is needed to establish the minimum level for detectable movement so that these key image quality issues can be evaluated.

Intervoxel Issues

Addressability. The number of voxels that are addressable is a function of the (x, y) image scanner bandwidth and the number of scanners used in a display system. The z-axis addressability is a function of the rotational speed of the projection surface and the modulation rate. Addressability limits define the image quality in that they determine the voxel spacing and overlap possible. The definition of surfaces and lines depends on adequate addressability. Often, the optics design requires a trade-off among addressability, laser spot size, luminance, and the number of voxels that can be displayed. Optimization of this trade-off is the key to providing useful volume displays.

Number of Voxels. The number of voxels that can be seen at any one time determines scene density in a volume display. A volume image consists of discrete emitters, reflectors, and occluders. The visual fidelity of volume objects depends on the total number of voxels available to sculpt that object. The number of points available is a function of the bandwith of the (x, y) scanner, the speed of rotation of the projection surface, the optical design, and the number of scanners. The number of voxels required is a function of the viewer's task, the object voxel definition, and the viewing distance. The key hurdle in the production of volume displays is achieving parity between the number of voxels required and the number of voxels available.

Temporal Display Issues

Flicker. Probably the most critical human factors requirement for multiplanar displays is the level of image flicker. This requirement determines the refresh rate of the image and is a major factor in the perceived image quality of volume displays. Also, with the potential for increased luminance using a laser, flicker thresholds increase according to the Ferry-Porter law, which states that flicker thresholds are related linearly to the logarithm of the stimulus luminance. The flicker problem could also be compounded by a large-area display device, which stimulates the peripheral areas of the retina, where the eye is more sensitive to flicker. There are large individual differences in the perception of flicker. The ANSI standard states that a display should be flicker-free for 90 percent of the viewers and provides a testing procedure. Multiplanar displays are, by nature, temporally limited, and the interaction of voxel refresh and perception must be examined in future research.

Other Temporal Effects. The time each voxel is turned on is a key system design parameter that influences several image quality measures. A laser beam of a given size when diffused on a projection surface can change in perceived area based on the duration of the voxel *on* time. A laser beam 2 mm in diameter appears to the eye to be much smaller when the *on* time is reduced. The *on* time determines the amount of time the eye has to integrate the point of light. The extremely low persistence of laser light on a rotating projection surface takes visual psychophysics-based display research into new frontiers. Much work is needed to understand and exploit the power of the temporal aspects of these systems.

Perceptual Display Issues

The availability of volume displays creates the need for research in several perceptual issues. The ability of the viewer to use parallax and motion parallax to understand the form of a volume object is the basic requirement for this technology.

Translucence and Form Perception. The degree of translucence in a volume image is determined by the multiplanar approach used to generate the image. All multiplanar images exhibit some transparency, and it is unlikely that any multiplanar display developed in the future will show opaque objects. There is a need to identify the tasks for which translucence is beneficial. Transparent images with high information content can be difficult to comprehend; information clutter precludes understanding. Transparency may be controllable by variations in color.

Surface Density. The perception of volume features such as form, relative position, orientation, and size depend on the number and spacing of the voxels in (x, y, z) space, or the surface density. The efficient use of scanner bandwidth requires use of the minimum number of voxels needed to achieve the perception desired. Conversely, critical tasks may require surface densities well above this minimum level.

Volume Icon Definition and Perception. Another critical design issue is how to use the limited number of voxels to sculpt volume icons. A similar problem was addressed by the original designers of the SpaceGraph system described in Chapter 11. Since the system could show only one point per plane, the question of how to locate the points to maintain maximum image clarity was critical. There has been some research to predict the level of definition needed for a given perceptual accuracy requirement. More needs to be done.

Modified ANSI Metrics. Image quality metrics are needed to ensure that a laser-based volume display meets minimum levels of acceptability to the viewers, and to provide a foundation for the superior image quality needed. Modified versions of the image quality specifications given in the ANSI standard are needed to extend and validate these metrics for volume displays. Extension of these metrics requires the addition of z dimension measurements and the parametric evaluation of the image quality from multiple viewing angles. These extended metrics should be provided for measuring luminance, contrast, jitter linearity, color contrast, resolution, and flicker. Two other important metrics are the following:

> *Modulation transfer function volume (MTFV)*: The display industry needs to conduct systematic linear systems analysis to develop a MTFV metric similar to the MTFA for 2D displays, which relates human visual system performance to display performance. This metric should consider spatial image characteristics, temporal characteristics, and color characteristics of the volume display.
>
> *Parametric view flicker test (PVFT)*: The ANSI measure for flicker can be used with an extension for parametrically changing viewing angle. The particular viewing angle of the observer can greatly influence the perception of flicker. A psychophysical model of temporal contrast sensitivity should be developed and verified with observers using the ANSI technique.

13.4 Future Technology for Volumetric Displays

Information display systems should be customized to provide the visualization needed to find the solution to a particular problem. Future display systems should therefore make use of a vast array of display technology to provide critical information in a natural and usable format to the viewers. The system designer of the future will be able to choose from a variety of display technologies to provide the basis for these custom display systems. Display options, once dominated by the home television market, have recently expanded dramatically. Alternatives now range in size from small wrist-mounted displays, to laptop flat panels, to large-group projection displays. Resolution and color capability have likewise seen dramatic improvement, providing additional options for the designer to meet the specific needs of the customer.

The use of 2D depth cues with standard flat-panel or CRT displays will provide important insights for many tasks. Stereoscopic displays have consistently improved over the years and will be an important option in the future for tasks requiring parallax cues of binocular disparity and binocular depth.

Volumetric display technology will provide an important option for team tasks and tasks requiring many simultaneous views of real-time or multidimensional data. Applications that may benefit immediately from volumetric display technology include air traffic control, computational fluid dynamics, weather pattern analysis, medical imaging, remote manipulation of objects, complex data interpretation, entertainment, education, and training for 3D volumetric tasks.

References

[ADEL91]   Adelson, S. J., Bentley, J. B., Chong, I. S., Hodges, L. F., and Winograd, J. 1991. Simultaneous generation of stereoscopic views. *Comput Graphics Forum,* 10:3–10.

[ADELn.d.]   Adelson, S. J., and Hodges, L. F. N.d. Ray-tracing stereoscopic images. *Vis Comput* (forthcoming).

[ANDE90]   Anderson, G. J. 1990. Focused attention in three-dimensional space. *Percept Psychophys* 47:112–20.

[APPL88]   Apple Computer, Inc. 1988. *Inside Macintosh.* Reading, MA: Addison-Wesley.

[ARDI86]   Arditi, A. 1986. Binocular vision. In *Handbook of perception and human performance,* vol. 1, ed. K. R. Boff, L. Kaufman, and J. P. Thomas. New York: Wiley.

[ARMS90]   Armstrong, A. W. P. 1990. Perspective and stereo for projection from and display of four dimensions. *Proc SPIE: Stereoscopic displays and applications* 1256:54–61.

[ARNO85]   Arnold, Steven M. 1985. Electron beam fabrication of computer-generated holograms. *Opt Eng* 24(5):803–7.

[BADC85]   Badcock, D. R., and Schor, C. M. 1985. Depth-increment detection function for individual spatial channels. *Opt Soc Am-A* 2:1211–16.

[BADL86]   Badler, N. I., Manoochehri, K. H., and Baraff, D. 1986. Multi-dimensional input techniques and articulated figure positioning by multiple constraints. 1986 Workshop on Interactive 3D Graphics, University of North Carolina, Chapel Hill, October 23–24.

[BADT88]   Badt, S., Jr. 1988. Two algorithms for taking advantage of temporal coherence in ray tracing. *Vis Comput* 4(3):123–32.

[BAKE87]   Baker, J. 1987. Generating images for a time-multiplexed stereoscopic computer graphics system. *Proc SPIE: True 3D imaging techniques and display technologies* 761:44–52.

[BARH90]   Barham, P. T., Harrison, L. L., and McAllister, D. F. 1990. Some variations on VISIDEP™ using computer graphics. *Opt Eng* 29:1504–6.

[BARH91]   Barham, P. T., and McAllister, D. F. 1991. Comparison of stereoscopic cursors for the interactive manipulation of B-splines. *Proc SPIE: Stereoscopic displays and applications II* 1457:18–26.

[BARN74]   Barnes, R. H., Moeller, C. E. Kircher, J. F., and Verber, C. M. 1974. Two-step excitation of flourescence in iodine monochloride vapor. *Appl Phys Lett* 24(12): 610–12.

[BARN91]   Bar-Natan, Dror. 1991. Random dot stereograms. *Mathematica* 1(3):69–75.

[BEAT86]   Beaton, R. J., DeHoff, R. J., and Knox, S. T. 1986. Revisiting the display flicker problem: Refresh rate requirements for field-sequential stereoscopic display systems. *Dig Tech Pap SID Int Symp* XVII:150.

[BEAT87]   Beaton, R. J., Dehoff, R. J., Weiman, N., and Hildebrandt, P. W. 1987. Evaluation of input devices for 3D computer display workstations. *Proc SPIE: True three-dimensional imaging techniques and display technologies* 761:94–101.

[BEAT88]   Beaton, R. J., and Weiman, W. 1988. User evaluation of cursor-positioning devices for 3-D display workstations. *Proc SPIE: Three-dimensional imaging and remote sensing imaging* 902:53–58.

[BEAT91]   Beaton, Robert J. 1991. On the image quality of 3D field sequential stereoscopic display systems. *SID Dig* 21:830.

[BENT69]   Benton, Stephen A. 1969. Hologram reconstructions with extended incoherent sources. *Opt Soc Am* 59:42–54.

[BENT80]   Benton, Stephen A. 1980. Holographic displays: 1975–1980. *Opt Eng* 19(5): 686–90.

[BENT82]   Benton, Stephen A. 1982. Survey of holographic stereograms. *Proc SPIE: Processing and display of three-dimensional data* 367:15–19.

[BENT87]   Benton, Stephen A. 1987. Alcove holograms for computer-aided design. *Proc SPIE: True 3D imaging techniques and display technologies* 761:53–61.

[BENT88]   Benton, Stephen A. 1988. Computer-graphic reflection alcove holograms. *Proc SPIE: Computer-generated holography II* 884.

[BERR50]   Berry, R. N., Riggs, L. A., and Duncan, C. P. 1950. The relation of vernier and depth discriminations to field brightness. *Exp Psychol* 40:349–54.

[BERR73]   Berreman, D. W. 1973. Optics in smoothly varying planar structures: Applications to liquid crystal twist cells. *J Opt Soc Am* 63:1374.

[BERR75]   Berreman, D. W. 1975. Liquid crystal twist cell dynamics with backflow. *J Appl Phys* 46(9):3746.

[BIER86]   Bier, E. A. 1986. Skitters and jacks: Interactive 3D positioning tools. 1986 Workshop on Interactive 3D Graphics, University of North Carolina, Chapel Hill, October 23–24.

[BLAK70]   Blakemore, C. 1970. The range and scope of binocular depth discrimination in man. *Physio* 211:599–622.

[BLAK87]   Blake, R., Wilson, H., and Pokorny, J. 1987. Diplopia thresholds and disparity scaling in the isolated blue cone system. *Suppl Invest Ophthalmol Vis Sci* (supp.) 28(3):293.

[BLAK89]   Blake, R. 1989. A neural theory of binocular rivalry. *Psychol Rev* 96:145–67.

[BOFF88]   Boff, K. R., and Lincoln, J. E. 1988. *Engineering data compendium: Human perception and performance*. Wright-Patterson AFB, Harry G. Armstrong Aerospace Medical Research Laboratory.

[BOS84]   Bos, P. J., and Koehler-Beran, K. R. 1984. The pi-cell: A fast liquid crystal optical switching device. *Mol Cryst Liq Cryst* 113:329.

[BOS89]   Bos, P. J., and Haven, T. 1989. Field-sequential stereoscopic viewing systems using passive glasses. *Proc Soc Inf Disp* 30(1):139.

[BOS93]   Bos, P. J. 1993. High contrast light shutter system. U.S. Patent 5,187,603.

[BOWN90]   Bowne, S. F., McKee, S. P., and Tyler, C. W. 1990. A disparity energy model of early stereo processing. *Suppl Invest Ophthalmol Vis Sci* 31(4):303.

[BÖRN87]   Börner, R. 1987. Progress in projection of parallax panoramagrams onto wide-angle lenticular screens. *Proc SPIE: True 3D imaging techniques and display technologies* 761:35–43.

[BREW56]   Brewster, David. 1856. *The stereoscope: Its history, theory and construction*. London: John Murray.

[BRID87]   Bridges, A. L., and Reising, J. M. 1987. Three-dimensional stereographic pictorial visual interfaces and display systems in flight simulation. *Proc SPIE: True 3D imaging techniques and display technologies* 761:102–12.

[BRIN83]   Brinkmann, U. 1983. A laser-based three-dimensional display. *Lasers Appl* 2(3):55–66.

[BROO89a]   Brookes, A., and Stevens, K. A. 1989. The analogy between stereo depth and brightness. *Perception* 18:601–14.

[BROO89b]   Brookes, A., and Stevens, K. A. 1989. Binocular depth from surfaces versus volumes. *Exp Psychol (Hum Percept)* 15:479–84.

[BULT87]   Bulthoff, H. H., and Mallot, H. A. 1987. Interaction of different modules in depth perception. *Suppl Invest Ophthalmol Vis Sci* 28(3):293.

[BURT88]   Burton, R. P., Becker, S. C., Broekhuijsen, B. J., Hale, B. J., and Richardson, A. E. 1988. Advanced concepts in device input for 3-D display. *Proc SPIE: Three-dimensional imaging and remote sensing imaging* 902:59–63.

[BUSQ90]   Busquets, A. M., Parrish, R. V., and Williams, S. P. 1990. Depth-viewing volume increase by collimation of stereo 3-D displays. IEEE Southeastcon '90, New Orleans, April.

[BUTT88a]   Butts, D. R. W., and McAllister, D. F. 1988. Implementation of true 3-D cursors in computer graphics. *Proc SPIE: Three-dimensional imaging and remote sensing imaging* 902:74–84.

[BUTT88b]   Butts, D. R. W. 1988. Implementation of an alternating-pair real-time stereoscopic cursor. Master's thesis, North Carolina State University, Raleigh.

[BUZA85]   Buzak, T. S. 1985. CRT displays full-color 3-D images. *Inf Disp* 1 (November): 12–13.

[BYAT81]   Byat, Dennis W. G. 1981. Stereoscopic television system. U.S. Patent 4,281,341.

[CARV91]   Carver, D. E., and McAllister, D. F. 1991. Development of stereoscopic 3-D drawing applications. *Proc SPIE: Stereoscopic displays and applications II* 1457:54–65.

[CAVA87]   Cavanagh, P. 1987. Reconstructing the third dimension: Interactions between color, texture, motion, binocular disparity and shape. *Computer Vision Graphics Image Process* 37:2.

[CHAN90]   Chang, J. J. 1990. New phenomena linking depth and luminance in stereoscopic motion. *Vision Res* 30:137–47.

[CHER88]   Chernicoff, S. 1988. *Macintosh revealed*. Vol. 2, *Programming with the toolbox*. 2d ed. Carmel, IN: Hayden.

[CHER90a]   Chernicoff, S. 1990. *Macintosh revealed*. Vol. 3, *Mastering the toolbox*. Carmel, IN: Hayden.

[CHER90b]   Chernicoff, S. 1990. *Macintosh revealed*. Vol. 1, *Unlocking the toolbox*. 2d ed. Carmel, IN: Hayden.

[CLAR83]   Clark, N., and Lagerwall, S. 1983. Ferroelectric liquid crystal electro-optics using the surface stabilized structure. *Mol Cryst Liq Cryst* 94:213.

[COLL67]   Collender, R. B. 1967. The stereoptiplexer: Competition for the hologram. *Inform Disp* 4(6):27–31.

[COND84]   Condon, C. 1984. An overview of three dimensional motion picture camera systems. *Proc SPIE: Optics in entertainment II* 462:38–40.

[CORM85]   Cormack, R. H., and Fox, R. 1985. The computation of disparity and depth in stereogram. *Percept Psychophys* 38:375–80.

[DEER92]   Deering, M. 1992. High resolution virtual reality. *Comput Graphics* 26(July):2.

[DEHO89]   DeHoff, R. J., and Hildebrandt, P. W. 1989. Cursor for use in 3-D imaging systems. U.S. Patent 4,808,979.

[DEVA91]   Devarajan, R., and McAllister, D. F. 1991. Ray tracing stereoscopic implicitly defined surfaces. *Proc SPIE: Stereoscopic displays and applications II* 1457: 37–48.

[DUWA81]   Duwaer, A. L., and Van Den Brink, G. 1981. What is the diplopia threshold? *Percept Psychophys,* 29:295–309.

[EICH90]   Eichenlaub, J. 1990. Autostereoscopic display for use with a personal computer. *Proc SPIE: Stereoscopic displays and applications* 1256:156–63.

[EPST77]   Epstein, W., ed. 1977. *Stability and constancy in visual perception.* New York: Wiley.

[FAHL88]   Fahle, M., and Westheimer, G. 1988. Local and global factors in disparity detection of rows of points. *Vision Res* 28:171–78.

[FEND67]   Fender, D. H., and Julesz, B. 1967. Extension of Panum's fusional area in binocularly stabilized vision. *J Opt Soc Am* 57:819–30.

[FEND83]   Fendick, M., and Westheimer, G. 1983. Effects of practice and the separation of test targets on foveal and peripheral stereoacuity. *Vision Res* 23:145–50.

[FERG80]   Fergason, J. L. 1980. Performance of a matrix display using surface mode. 1980 Biennial Display Research Conference, p. 177.

[FERG85]   Fergason, J. L. 1985. Polymer encapsulated nematic liquid crystals for display and light control applications. *Dig SID Int Symp Tech Pap* XVI:68.

[FERW82]   Ferwerda, J. G. 1982. *The world of 3-D: A practical guide to stereo photography.* Netherlands Society for Stereo Photography.

[FOLE90]   Foley, J. D., van Dam, A., Feiner, S. K., and Hughes, J. F. 1990. *Computer graphics: Principles and practice.* 2d ed. Reading, MA: Addison-Wesley.

[FORN87]   Fornaro, R. J., Hodges, L. F., McAllister, D. F., and Robbins, W. E. 1987. Three-dimensional geomorphic data display station (3DGDDS). NCSU Computer Studies Technical Report TR-15-87.

[FREE90]   Freeman, R. D., and Ohzawa, I. 1990. On the neurophysiological organization of binocular vision. *Vision Res* 30:1661–76.

[FREY83]   Freyer, J. L., Rerimutter, R. J., and Goodman, J. W. 1983. Digital holography: Algorithms, E-beam lithography, and 3-D display. *Proc SPIE: International conference on computer-generated holography* 437:38–47.

[FUCH82]   Fuches, Henry, Pizer, S. M., Heinz, E. R., Bloomberg, S. H., Tsai, L. C., and Strickland, D. C. 1982. Design of an image editor with a space-filling three-dimensional display based on a standard raster graphic system. *Proc SPIE: Processing and display of three-dimensional data* 367:117–27.

[FUSE80]   Fusek, R. L., and Huff, L. 1980. Use of a holographic lens for producing cylindrical holographic stereograms. *Proc SPIE: Recent advances in holography* 215: 32–38.

[GABO48]   Gabor, D. 1948. A new microscopic principle. *Nature* 161:777–79.

[GERV88]   Gervautz, M., and Purgathofer, G. 1988. A simple method for color quantization: Octree quantization. In *New trends in computer graphics,* ed. N. Magenenat-Thalmann and D. Thalmann, pp. 219–31. New York: Springer-Verlag.

[GESC90]   Gescholm, M. Y. 1990. Private communication.

[GETT86]   Getty, D. J., and Huggins, A. W. F. 1986. Volumetric 3-D displays and spatial perception. In *Statistical image processing and graphics,* ed. E. J. Wegman and D. J. DePriest, pp. 321–43. New York: Marcel Dekker.

[GILL88]   Gillam, B., Chambers, D., and Russo, T. 1988. Postfusional latency in stereoscopic slant perception and the primitives of stereopsis. *Exp Psychol (Hum Percept)* 14:163–75.

[GOGE90]   Gogel, W. C. 1990. A theory of phenomenal geometry and its applications. *Percept Psychophys* 48(2):105–23.

[GOLD89]   Goldstein, E. B. 1989. *Sensation and perception.* Belmont, CA: Wadsworth.

[GOOC75]   Gooch, C. H., and Tarry, H. A. 1975. The optical properties of twisted nematic liquid crystal structure with twist angles $\leq 90°$. *J Appl Phys* 8:1575–84.

[GROT83]   Grotch, S. L. 1983. Three-dimensional and stereoscopic graphics for scientific data display and analysis. *IEEE Comput Graphics Appl* 3(8):31–43.

[GRUN86]   Grunwald, A. J., Ellis, S. R. and Smith, S. 1986. Spatial orientation from pictorial perspective displays. Proceedings of the 22nd Annual Conference on Manual Control, Dayton, Ohio, July.

[GRUN88]   Grunwald, A. J. and Ellis, S. R. 1988. Spatial orientation by familiarity cues. In *Training, human decision making and control,* ed. J. Patrick and K. D. Duncan. North-Holland, NY: Elsevier Science Publishers.

[HALP87]   Halpern, D. L., Patterson, R., and Blake, R. 1987. What causes stereoscopic tilt from spatial frequency disparity. *Vision Res* 27:1619–29.

[HARR86]   Harris, M. R., Geddes, A. J., and North, A. C. T. 1986. Frame-sequential stereoscopic system for use in television and computer graphics. *Displays* 7(1):12.

[HART87]   Hartman, W. J. A. M., and Hikspoors, H. M. J. 1987. Three-dimensional TV with cordless FLC spectacles. *Inform Disp* 3(9):15.

[HATT90]   Hattori, T. 1990. Three-dimensional photographing and three-dimensional playback device by spatial time-sharing method. U.S. Patent 4,943,860.

[HAVE87]   Haven, T. J. 1987. A liquid-crystal video stereoscope with high extinction ratios, a 28% transmission state, and one-hundred-microsecond switching. *Proc SPIE: True 3D imaging techniques and display technologies* 761:23–26.

[HEBB91]   Hebbar, P. D., and McAllister, D. F. 1991. Color quantization aspects in stereopsis. *Proc SPIE: Stereoscopic displays and applications II* 1457:233–41.

[HECK82]   Heckbert, P. 1982. Color image quantization for frame buffer display. *Comput Graphics* 16(3):297–307.

[HELM25]   Helmholtz, H. L. F. 1925. *Helmholtz's treatise on physiological optics, part III.* Rochester, NY: Optical Society of America.

[HILL90]   Hill, F. S. 1990. *Computer graphics.* New York: Macmillan.

[HOBG69]   Hobgood, W. S. 1969. A three-dimensional computer graphics display using a varifocal mirror. Master's thesis, University of North Carolina, Chapel Hill.

[HODG85]   Hodges, L. F., and McAllister, D. F. 1985. Stereo and alternating-pair techniques for display of computer-generated images. *IEEE Comput Graphics Appl* 5(9): 38–45.

[HODG87]   Hodges, L. F., and McAllister, D. F. 1987. True three-dimensional CRT-based displays. *Inform Disp* 3(5):18–22.

[HODG88a]   Hodges, L. F. 1988. Computing stereoscopic images. *J Theoret Graphics Comput* 1(1):1–10.

[HODG88b]   Hodges, L. F., and McAllister, D. F. 1988. Chromostereoscopic CRT-based display. *Proc SPIE: Three-dimensional imaging and remote sensing imaging* 902: 37–44.

[HODG88c]   Hodges, L. F. 1988. Technologies, applications, hardcopy and perspective transformation for true three-dimensional CRT-based display. Ph.D. diss., North Carolina State University.

[HODG92]   Hodges, L. F. 1992. Time-multiplexed stereoscopic computer graphics. *IEEE Comput Graphics Appl* 12(2):20–30.

[HONG89]   Hong, X., and Regan, D. 1989. Visual field defects for unidirectional and oscillatory motion in depth. *Vision Res* 29:809–19.

[HOUL86]   Houle, G., and Dubois, E. 1986. Quantization of color images for display on graphics terminals. In *GLOBECOM '86: IEEE Global Telecommunications Conference,* vol. 2, pp. 1138–42. Piscataway, NJ: IEEE Press.

[HUBB81]   Hubbard, R. L., and Bos, P. J. 1981. Optical-bounce removal and turnoff-time reduction in twisted-nematic displays. *IEEE Trans Electron Devices* 28(6): 723.

[HULL21]   Hull, L. M. 1921. The cathode-ray oscillograph and its application in radio work. *Proceedings of the IRE,* vol. 9, pp. 130–49.

[INOU90]   Inoue, T., and Ohzu, H. 1990. Measurement of the human factors of 3-D images on a large screen. *Proc SPIE: Large-screen projection displays II* 1255:104–7.

[ITTE60]   Ittelson, W. H. 1960. *Visual space perception.* New York: Springer.

[IVES03]   Ives, F. E. 1903. Parallax stereogram and process of making same. U.S. Patent 725,567.

[IVES28]   Ives, H. E. 1928. A camera for making parallel panoramagrams. *J Opt Soc Amer* 17(6):435–39.

[JENT14]   Jentzsch, F. 1914. The binocular microscope. *J Royal Microscopical Soc* 1914: 1–16.

[JERR48]   Jerrard, H. J. 1948. Optical compensators for measurement of elliptical polarization. *J Opt Soc Am* 38(1):35–59.

[JONE84]   Jones, E. R., Jr., McLaurin, A. P., and Cathey, L. 1984. VISIDEP™: Visual image depth enhancement by parallax induction. *Proc SPIE: Advances in display technology IV* 457:16–19.

[JONE85]   Jones, E. R. 1985. Three dimensional video apparatus and methods using composite and mixed images. U.S. Patent 4,528,587.

[JONE90]   Jones, E. R., and McLaurin, A. P. 1990. Stereoscopic medical viewing device. U.S. Patent 4,924,853.

[JORD90]   Jordan, J. R. III., Geisler, W. S., and Bovik, A. C. 1990. Color as a source of information in the stereo correspondence problem. *Vision Res* 30:1955–70.

[JULE71]   Julesz, B. 1971. *Foundations of cyclopean perception*. Chicago: University of Chicago Press.

[JULE81]   Julesz, B. 1981. Textons, the elements of texture perception, and their interactions. *Nature* 290:91–97.

[JULE83]   Julesz, B. 1983 Texton theory of two-dimensional and three-dimensional vision. *Proc SPIE: Processing and display of three-dimensional data* 367:2.

[KANE87]   Kaneko, E. 1987. *Liquid crystal TV displays: Principles and applications of liquid crystal displays*. Tokyo: KTC Scientific Publishers.

[KAUF74]   Kaufman, Lloyd. 1974. *Sight and mind: An introduction to visual perception*. New York: Oxford University Press.

[KELL89]   Kelly, D. H. 1989. Retinal inhomogeneity and motion in depth. *Opt Soc Am A* 6:98–105.

[KILP76]   Kilpatrick, P. J. 1976. The use of a kinesthetic supplement in an interactive graphics system. Ph.D. diss., University of North Carolina, Chapel Hill.

[KIM87]   Kim, W. S., Ellis, S. R., Tyler, M. E., Hannaford, B., and Stark, L. W. 1987. Quantitative evaluation of perspective and stereoscopic displays in three-axis manual tracking tasks, *IEEE Trans Syst Man Cybern* SMC-17 (1):61–72.

[KISH65]   Kishto, B. N. 1965. The color stereoscopic effect. *Vision Res* 5(June):313–29.

[KOLL88]   Kollin, J. S. 1988. Collimated view multiplexing: A new approach to 3-D. *Proc SPIE: Three-dimensional imaging and remote sensing imaging* 902:24–30.

[KWON79]   Kwon, Y. S. Lefkowitz, I., and Lontz, R. 1979. Improved liquid crystal device response time. *Appl Opt* 18(11):1700.

[LAND40]   Land, E. H. 1940. Vectographs: Images in terms of vectorial inequalities and their application in three-dimensional representation. *J Opt Soc Am* 30:230.

[LAND91]   Landy, M. S., Maloney, L. T., and Young, M. J. 1991. Psychophysical estimation of the human depth combination rule. *Proc SPIE: Sensor fusion III: 3D perception and recognition* 1383:247–54.

[LANE82]   Lane, Bruce. 1982. Stereoscopic displays. *Proc SPIE: Processing and display of three-dimensional data* 367:20–32.

[LATH79]   Lathrop, Irvin T., and Kunst, Robert J. 1979. *Photo-offset*. Chicago: American Technical Society.

[LEGG89]   Legge, G., and Gu, Y. 1989. Stereopsis and contrast. *Vision Res* 29:989–1004.

[LESL70]   Leslie, T. M. 1970. Distortion of twisted orientation patterns in liquid crystals by magnetic fields. *Mol Cryst Liq Cryst* 12:57–72.

[LEVE68]   Levelt, W. J. M. 1968. On binocular rivalry. The Hague: Mouton.

[LEVI91]   Levit, C. C., and Bryson, S. 1991. A virtual environment for the interactive exploration of 3-D steady flows. *Proc SPIE: Stereoscopic displays and applications II* 1457:161–68.

[LEVK84]   Levkowitz, H., Trivedi, S. S., and Udupa, J. K. 1984. Interactive manipulation of 3D data via a 2D display device. *Proc SPIE: Processing and display of three-dimensional data II* 507:25–37.

[LEWI71]   Lewis, Jordan D., Verber, Carl M., and McGhee, A. 1971. True three-dimensional display. *IEEE Trans Electron Devices* ED-18(9):724–32.

[LIPS79]   Lipscomb, J. 1979. Three-dimensional cues for a molecular computer graphics system. Ph.D. diss. University of North Carolina, Chapel Hill.

[LIPS89]   Lipscomb, J. 1989. Experience with steroscopic display devices and output algorithms. *Proc SPIE: Non-Holographic True 3D Display Techniques* 1083:28–34.

[LIPT patent]   Lipton, L., Berman, A., and Meyer, L. Achromatic liquid crystal shutter for stereoscopic and other applications. U.S. Patent 4,884,876.

[LIPT82]   Lipton, Lenny. 1982. *Foundations of the stereoscopic cinema*. New York: Van Nostrand Reinhold.

[LIPT84a]   Lipton, Lenny, and Meyer, L. 1984. A time-multiplexed two-times vertical frequency stereoscope video system. *Dig SID Symp Tech Pap* XV.

[LIPT84b]   Lipton, Lenny. 1984. Binocular symmetries as criteria for the successful transmission of images. *Proc SPIE: Processing and display of three-dimensional data II* 507:108–13.

[LIPT87]   Lipton, Lenny. 1987. Factors affecting ghosting in a time-multiplexed planostereoscopic CRT display system. *Proc SPIE: True 3D imaging techniques and display technologies* 761:75–78.

[LIPT88]   Lipton, L., and Berman, A. 1988. Push-pull liquid crystal modulator for electronic stereoscopic display. *Proc SPIE: Three-dimensional imaging and remote sensing imaging* 902:31–36.

[LIPT89]   Lipton, L. 1989. Compatability of stereoscopic video systems with broadcast television standards. *Proc SPIE: Three-dimensional visualization and display technologies* 1083:95–101.

[LIPT90]   Lipton, L., and Ackerman, M. 1990. Liquid crystal shutter system for stereoscopic applications. U.S. Patent 4,967,268.

[LIVI88]   Livingstone, M., and Hubel, D. 1988. Segregation of form, color, movement, and depth: Anatomy, physiology, and perception. *Science* 240:740–49.

[LOGO90]   Logothetis, N. K., Schiller, P. H., Charles, E. R., and Hurlbert, C. 1990. Perceptual deficits and the activity of the color-opponent and broad-band pathways at isoluminance. *Science* 247:214–17.

[LOVE90]   Love, S. 1990. Nonholographic, autostereoscopic, nonplanar display of computer generated images. Ph.D. diss., North Carolina State University.

[MACA54]   MacAdam, D. L. 1954. Stereoscopic perceptions of size, shape, distance and direction. *SMPTE* 62:271–93.

[MANS90]   Mansfield, J. S., and Parker, A. J. 1990. Disparity averaging in bandpass-filtered stereograms. *Suppl Invest Ophthalmol Vis Sci* 31(4):96.

[MARR82]   Marr, D. 1982. *Vision*. San Francisco: W. H. Freeman, p. 129.

[MARS84]   Marshall, Grayson, and Gundlach, Gregory. 1984. Method of making a three-dimensional photograph. U.S. Patent 4,481,050.

[MASH86]   Mash, D. H., Crossland, W. A., and Morrissy, J. H. 1986. Improvements in or relating to stereoscopic display device. UK Patent 1,448,520.

[MAYH87]   Mayhew, C. A. 1987. True three-dimensional animation in motion pictures. *Proc SPIE: True 3D imaging techniques and display technologies* 761:133–35.

[MAYH90]   Mayhew, C. A. Texture and depth enhancement for motion pictures and television. *SMPTE* 99 (October): 809–14.

[MCGR84]   McGreevy, M. W., and Ellis, S. R. 1984. Direction judgment errors in perspective displays. 20th Annual Conference on Manual Control, Ames Research Center, Moffett Field, CA, June 12–14.

[MCGR85]   McGreevy, M. W., Ratzlaff, C. R., and Ellis, S. R. 1985. Virtual space and two-dimensional effects in perspective displays. 21st Annual Conference on Manual Control, NASA CP-2428, June.

[MCKE83]   McKee, S. P. 1983. The spatial requirements for fine stereoacuity. *Vision Res* 23:191–98.

[MCKE88]   McKee, S. P., and Mitchison, G. J. 1988. The role of retinal correspondence in stereoscopic matching. *Vision Res* 28:1001–12.

[MCKE90]   McKee, S. P., Levi, D., and Bowne, S. F. 1990. The imprecision of stereopsis. *Vision Res* 30:1763–79.

[MCLA85]   McLaurin, A. P., Jones, E. R., Jr., and Cathey, L. 1985. Electronic three-dimensional vision for mechanical devices through VISIDEP™. In *Vision '85 conference proceedings*. Dearborn, MI: Machine Vision Association of the Society of Manufacturing Engineers, pp. 464–69.

[MCLA86]   McLaurin, A. P., Jones, E. R., Jr., and Cathey, L. 1986. Visual image depth enhancement process: An approach to three-dimesional imaging. *Displays* 7: 111–15.

[MCLA87]   McLaurin, A. P., Jones, E. R., Jr., and Cathey, L. 1987. Single-source three-dimensional imaging system for remote sensing. *Opt Eng* 26:1251–56.

[MCLA89]   McLaurin, A. P., Jones, E. R., Jr., and Cathey, L. 1989. Three-dimensional display methods apparatus. U.S. Patent 4,807,024.

[MCLA90]   McLaurin, A. P., Jones, E. R., Jr., and Mason, J. Lorin, Jr. 1990. Three-dimensional endoscopy through alternating-frame technology. *Proc SPIE: Stereoscopic displays and applications* 1256: 307–11.

[MILL86]   Millgram, P., and Van der Horst, R. 1986. Alternating-field stereoscopic displays using light scattering liquid crystal spectacles. *Displays* 7(2):67–72.

[MITC66a]   Mitchell, D. E. 1966. A review of the concept of "Panum's fusional areas." *Am J Optom* 43:387–401.

[MITC66b]   Mitchell, D. E. 1966. Retinal disparity and diplopia. *Vision Res* 6:441–51.

[MITC84]   Mitchison, G. J., and Westheimer, G. 1984. The perception of depth in simple figures. *Vision Res* 24:1063–73.

[MITC87]   Mitchison, G. J., and McKee, S. P. 1987. The resolution of ambiguous stereoscopic matches by interpolation. *Vision Res* 27:285–94.

[MITC88]   Mitchison, G. 1988. Planarity and segmentation in stereoscopic matching. *Perception* 17:753–82.

[MOLT84]   Molteni, W. J. 1984. Natural color holographic stereograms by superimposing three rainbow holograms. *Proc SPIE: Optics in entertainment II* 462:14–19.

[MORG82]   Morgan, H., and Symmes, D. 1982. *Amazing 3D*. Boston: Little, Brown.

[MOSH88]   Mosher, C. E., Jr., Sherouse, F. W., Chaney, E. L., and Rosenman, J. G. 1988. 3-D displays and user interface design for a radiation therapy treatment planning CAD tool. *Proc SPIE: Three-dimensional imaging and remote sensing imaging* 902:64–72.

[MUIR61]   Muirhead, J. C. 1961. Variable focus length mirrors. *Rev Sci Instrum* 32:210.

[MURC82]   Murch, G. M. 1982. Visual accommodation and convergence to multichromatic information displays. *Dig SID Int Symp Tech Pap* XIII:192–93.

[MUST85]   Mustillo, P. 1985. Binocular mechanisms mediating crossed and uncrossed stereopsis. *Psychol Bull* 97:187–201.

[NAKA86]   Nakayama, K., and Silverman, G. H. 1986. Serial and parallel processing of visual feature conjunctions. *Nature* 320:264–65.

[NAKA89]   Nakayama, F., Shimojo, S., and Silverman, G. H. 1989. Stereoscopic depth: Its relation to image segmentation, grouping, and the recognition of occluded objects. *Perception* 18(1):55–68.

[NIEL86]   Nielson, G. M., and Olsen, Dan R., Jr. 1986. Direct manipulation techniques for 3D objects using 2D locator devices. 1986 Workshop on Interactive 3D Graphics, University of North Carolina, Chapel Hill, October 23–24.

[NOBL87]   Noble, L. 1987. Use of lenses to enhance depth perception. *Proc SPIE: True 3D imaging techniques and display technologies* 761:126–28.

[OBRI88]   O' Brien, T. P. 1988. Method for displaying three-dimensional images using composite driving waveforms. U.S. Patent 4,747,665.

[OGLE63]   Ogle, K. N. 1963. Stereoscopic depth perception and exposure delay between images to the two eyes. *J Opt Soc Am* 53:1296.

[OKOS76]   Okoshi, T. 1976. *Three-dimensional imaging techniques*. New York: Academic Press.

[ONO88]   Ono, H., Rogers, B., Ohmi, M., and Ono, M. E. 1988. Dynamic occlusion and motion parallax in depth perception. *Perception* 17:255–66.

[OWEN75] Owens, D. A., and Leibowitz, A. W. 1975. Chromostereopsis with small pupils. *J Opt Soc Am* 65:358–59.

[PAPA87]   Papathomas, T. V., Schiavone, J. A., and Julesz, B. 1987. Stereo animation for very large data bases: Case study—meteorology. *IEEE Comput Graphics Appli* 7:18–27.

[PARR90a] Parrish, R. V., and Williams, S. P. 1990. Determination of depth-viewing volumes for stereo three-dimensional graphic displays. NASA Technical Paper 2999, June.

[PARR90b] Parrish, R. V., Busquets, A. M., and Williams, S. P. 1990. Recent research results in stereo 3-D pictorial displays at Langley Research Center. 9th Digital Avionics Systems Conference, Virginia Beach, VA, October.

[PATT84]   Patterson, R., and Fox, R. 1984. The effect of testing method on stereoanomaly. *Vision Res* 24:403–08.

[PATT92a] Patterson, R., and Martin, W. 1992. Human stereopsis. *Hum Factors* 34:669–92.

[PATT92b]   Patterson, R., Moe, L., and Hewitt, T. 1992. Factors that affect depth perception in stereoscopic display. *Hum Factors* 34:655–68.

[PENN86]   Penna, M. A., and Patterson, R. 1986. *Projective geometry and its applications to computer graphics*. Englewood Cliffs, NJ: Prentice-Hall.

[PEPP81]   Pepper, R. L., Smith, D. C., and Cole, R. E. 1981. Stereo TV improves operator performance under degraded visibility conditions. *Opt Eng* 20:579–85.

[POGG88]   Poggio, G. F., Gonzalez, F., and Krause, F. 1988. Stereoscopic mechanisms in monkey visual cortex: Binocular correlation and disparity selectivity. *J Neurosci* 8:4531–50.

[PRIE75]   Priestly, E. B., Wojtowicz, P. J., and Sheng, P., eds. 1975. *Introduction to liquid crystals*. New York: Plenum Press.

[PRIN patent] Prince, D. W. IR light self synchronizing liquid crystal sound driver. Patent pending.

[PULF22]   Pulfrich, C. 1922. Die stereoskopie im dienste der isochromen und heterochromen photometrie. *Naturewissenschaften* 10:533–64, 569–601, 714–22, 735–43, 751–61.

[RAWS68]   Rawson, E. G. 1968. 3-D computer-generated movies using a varifocal mirror, *Appl Opt* 7:1505.

[RAWS69]   Rawson, E. G. 1969. Vibrating varifocal mirrors for 3-D imaging. *IEEE Spectrum* 6(9):37–43.

[RAYN74]   Raynes, E. P., and Shanks, I. A. 1974. Fast switching twisted nematic electro-optical shutter and color filter. *Electron Lett* 4:114.

[REGA73]   Regan, D., and Beverley, K. I. 1973. Some dynamic features of depth perception. *Vision Res* 13:2369–79.

[REGA86]   Regan, D., Erkelenst, C. J., and Collewijn, H. 1986. Necessary conditions for the perception of motion in depth. *Invest Ophthalmol Vis Sci* 27:584–97.

[REIN90]   Reinhart, W. F., Beaton, R. J., and Snyder, H. L. 1990. Comparison of depth cues for relative depth judgements. *Proc SPIE: Stereoscopic displays and applications* 1256:12–21.

[RICH70]   Richards, W. 1970. Stereopsis and stereoblindness. *Exp Brain Res* 10:380–88.

[ROBI91]   Robinette, W., and Rollcod, J. P. 1991. A computational model for the stereoscopic optics of a head mounted display. *Proc SPIE: Stereoscopic displays and applications II* 1457:140–60.

[ROCK85]   Rock, I., 1985. *The logic of perception*. Cambridge, MA: MIT Press.

[ROES76]   Roese, J. A., and Khalafalla, A. S. 1976. Stereoscopic viewing with PLZT ceramics. *Ferroelectrics* 10:47.

[ROES77]   Roese, J. A. 1977. Liquid crystal stereoscopic viewer. U.S. Patent 4,021,846.

[ROES78]   Roese, J. A., and McCleary, L. E. 1978. Stereoscopic computer graphics using PLZT electro-optic ceramics. *Proc Soc Inf Disp* 19(2):69–73.

[ROES79]   Roese, J. A., and McCleary, L. E. 1979. Stereoscopic computer graphics for simulation and modeling. *Proc SIGGRAPH* 13(2):41–47.

[ROES87]   Roese, J. A. 1987. Stereoscopic electro-optic shutter CRT displays—a basic approach. *Proc SPIE: Processing and display of three-dimensional data II* 507: 102–07.

[ROGE82]   Rogers, B., and Graham, M. 1982. Similarities between motion parallax and stereopsis in human depth perception. *Vision Res* 22:261–70.

[ROGE89]   Rogers, B., and Cagenello, B. 1989. Disparity curvature and the perception of three-dimensional surfaces. *Nature* 339:135–37.

[ROGE90]   Rogers, David F., and Adams, J. Alan. 1990. *Mathematical elements for computer graphics.* 2d ed. New York: McGraw-Hill.

[ROHA90]   Rohaly, A., and Wilson, H. R. 1990. Contrasts on binocular fusion do not depend on temporal frequency or color. *Suppl Invest Ophthamol Vis Sci* 31(4):304.

[SAND89]   Sandin, D., Sandor, E., Cunnally, W., Resch, M., DeFanti, T., and Brown, M. 1989. Computer-generated barrier-strip autostereography. *Proc SPIE: Nonholographic true three-dimensional display technologies* 1083:65–75.

[SAND90]   Sandin, D., Sandor, E., Cunnally, W., Meyers, S., DeFanti, T., and Brown, M. 1990. Computer-interleaved barrier strip autostereograms: proofing/mass-printing advances and cylindrical projections. *Proc SPIE: Stereoscopic displays and applications* 1256:312–21.

[SATO77]   Sato, T., and Wada, M. 1977. Reduction of response and recovery times of nematic cells with electrically controlled birefringence. *Electron Commun Jpn* 60-C (11):128.

[SATO90]   Sato, T., Nishimura, K., and Ohzawa, I. 1990. Visual evoked potentials to equiluminant dynamic random-dot stereogram. *Suppl Invest Ophthalmol Vis Sci* 31(4):92.

[SAUN68]   Saunders, B. G. 1968. Stereoscopic drawing by computer—is it orthoscopic? *Appl Opt* 7(8):1499–1503.

[SCHI90]   Schiller, P. H., Logothetis, N. K., and Charles, E. R. 1990. Functions of the colour-opponent and broad-band channels of the visual system. *Nature* 343:68–70.

[SCHM83]   Schmandt, C. 1983. Spatial input/display correspondence in a stereoscopic computer graphic work station. *Comput Graphics,* 17(3):253–61.

[SCHO81]   Schor, C. M., and Tyler, C. W. 1981. Spatio-temporal properties of Panum's fusional area. *Vision Res* 21:683–92.

[SCHO88]   Schor, C. M. 1988. Imbalanced adaptation of accommodation and vergence produces opposite extremes of the AC/A and CA/C ratios. *Am J Optom Physiol Opt* 65:341–48.

[SCHO91]   Schor, C. M. 1991. Spatial constraints of stereopsis in video displays. In *Pictorial communication,* ed. S. Ellis, M. Kaiser, and A. Grunwald. Bristol, PA: Taylor and Francis.

[SCHU79]   Schumer, R., and Ganz, L. 1979. Independent stereoscopic channels for different extents of spatial pooling. *Vision Res* 19:1303–14.

[SEDG86]   Sedgwick, H. A. 1986. Space perception. In *Handbook of perception and human performance,* vol. 1, ed. K. Boff, L. Kaufman, and J. P. Thomas, pp. 1–57. New York: Wiley.

[SEXT89]   Sexton, I. 1989. Parallax barrier 3DTV. *Proc SPIE: Three-dimensional visualization and display technologies* 1083:84–94.

[SHEA82]   Shea, S. L., Feustel, T. C., and Aslin, R. N. 1982. Cyclopean micropia: Scanning stereoscopic forms. *Suppl Invest Ophthalmol Vis Sci* 23(3):267.

[SHEA90]   Shea, S. L. 1990. Accommodative changes to random element stereogram. *Suppl Invest Ophthalmol Vis Sci* 31(4):94.

[SHER78]   Sher, L. D. 1978. Three-dimensional display. U.S. Patent 4,130,832.

[SHER85]   Sher, L. D., and Barry, C. D. 1985. The use of an oscillating mirror for three-dimensional displays. In *New methodologies in studies of protein configurations,* ed. T. T. Wu, pp. 165–89. New York: Van Norstrand Reinhold.

[SHIM88]   Shimojo, S., Silverman, G., and Nakayama, K. 1988. An occlusion-related mechanism of depth perception based on motion and interocular sequence. *Nature* 333(6170):265–67.

[SHUR64]   Shurcliff, W. A., and Ballard, S. T. 1964. *Polarized light*. Princeton: D. Van Nostrand.

[SIST89]   Sistare, S., and Friedell, M. 1989. A distributed system for near-real-time display of shaded three-dimensional graphics. Proceedings, Graphics Interface '89, pp. 283–90.

[SPOT53]   Spottiswoode, Raymond, and Spottiswoode, Nigel. 1953. *The theory of stereoscopic transmission and its application to the motion picture*. Berkeley: University of California Press.

[STAR90]   Starks, M. 1990. Portable low-cost devices for videotaping, editing, and displaying field sequential stereoscopic motion pictures in video. *Proc SPIE: Stereoscopic displays and applications* 1256:266–69.

[STAR91]   Starks, Michael. 1991. Stereoscopic video and the quest for virtual reality: An annotated bibliography of selected topics. *Proc SPIE: Stereoscopic displays and applications II* 1457:327–37.

[STEE87]   Steenblik, R. A. 1987. The chromostereoscopic process: A novel single image stereoscopic process. *Proc SPIE: True 3D imaging techniques and display technologies* 761:27–34.

[STEE89]   Steenblik, R. A. 1989. Chromostereoscopic microscopy. *Proc SPIE: Three dimensional visualization and display technologies* 1083:60–64.

[STEV88]   Stevens, K. A., and Brookes, A. 1988. Integrating stereopsis with monocular interpretations of planar surfaces. *Vision Res* 28:371–86.

[STEV90]   Stevens, K. A. 1990. Constructing the perception of surfaces from multiple cues. *Mind and Language* 5:253–66.

[STEV92]   Stevenson, S. B., Cormack, L. K., Schor, C. M., and Tyler, C. W. 1992. Disparity tuning in mechanisms of human stereopsis. *Vision Res* 32:1685–94.

[STRO70]   Stromeyer, C. F., and Psotka, J. 1970. The detailed texture of eidetic images. *Nature* 225:346.

[SUN88]   Sun Microsystems, Inc. 1988. *TAAC-1 Application Accelerator: Software reference manual*, rev. A.

[SUND72]   Sundet, J. M. 1972. The effect of pupil size variation on the colour stereoscopic phenomenon. *Vision Res* 12(May):1027–32.

[TEIC85]   Teicholz, E. 1985. *CAD.CAM handbooks*. New York: McGraw-Hill, pp. 464–68.

[TILT71]   Tilton, H. B. 1971. Real-time direct-viewed CRT displays having holographic properties. Proceedings of the Technical Program, Electro-Optical Systems Design Conference, 1971 West, Anaheim, CA, May 18–20, pp. 415–22.

[TILT77]   Tilton, H. B. 1977. An autostereoscopic CRT display. *Proc SPIE: Three-dimensional imaging* 120:68–72.

[TILT82]   Tilton, H. B. 1982. Holoform oscillography with a parallactiscope. *Dig SID Int Symp Tech Pap* XIII: 276–77.

[TILT85]   Tilton, H. B. 1985. Large-CRT holoform display. Conference Record of the 1985 International Display Research Conference, Society for Information Display, Playa del Rey, CA, pp. 145–46.

[TILT87]   Tilton, H. B. 1987. The 3-D oscilloscope: A practical manual and guide. Englewood Cliffs, NJ: Prentice-Hall.

[TILT88]   Tilton, H. B. 1988. Nineteen-inch parallactiscope. *Proc SPIE: Three-dimensional imaging and remote sensing imaging* 902:17–23.

[TILT89]   Tilton, H. B. 1989. Everyman's real-time real 3-D. *Proc SPIE: Three-dimensional visualization and display technologies* 1083:76–83.

[TITT88]   Tittle, J. S., Rouse, M. W., and Braunstein, M. L. 1988. Relationship of static stereoscopic depth perception to performance with dynamic stereoscopic displays. In *Proceedings of the Human Factors Society 32nd Annual Meeting,* vol. 2, pp. 1439–42. Santa Monica, CA: Human Factors Society.

[TRAU67a]   Traub, A. C. 1967. Two teaching demonstrations using flexible mirrors. *Amer J Phys* 35:534–5.

[TRAU67b]   Traub, A. C. 1967. Stereoscopic display using rapid varifocal mirror oscillations. *Appl Opt* 6:1085–7.

[TRAU68]   Traub, A. C. 1968. A new 3-dimensional display technique. Report M68-4 of the Mitre Corp.

[TRAU70]   Traub, A. C. 1970. Three-dimensional display. U.S. Patent 3,493,290.

[TRAVn.d.]   Travis, A. R. L. N.d. Autostereoscopic 3-D display. *Appl Opt* (forthcoming).

[TYLE77]   Tyler, C. W. 1977. Spatial limits of human stereoscopic vision. *Proc SPIE: Three-dimensional imaging* 120:36–42.

[TYLE90a]   Tyler, C. W., and Clarke, M. B. 1990. The autostereogram. *Proc SPIE: Stereoscopic displays and applications* 1256:187.

[TYLE90b]   Tyler, C. W. 1990. A stereoscopic view of visual processing streams. *Vision Res* 30:1877–95.

[TYRE90]   Tyrell, R. A., and Leibowitz, H. W. 1990. The relation of vergence effort to reports of visual fatigue following prolonged near work. *Hum Factors* 32: 341–58.

[UEHA75]   Uehara, K., Mada, H., and Kobayashi, S. 1975. Reduction of electro-optical response times of a field-effect liquid crystal device. *IEEE Trans Electron Devices* 22:804.

[UPAT80]   Upatnieks, J., and Embach, J. T. 1980. 360-degree hologram displays. *Opt Eng* 19(5):696–704.

[VALY62]   Valyus, N. A. 1962. *Stereoscopy.* London, New York: Focal Press.

[VAND75]   Van Doorn, C. Z. 1975. Transient behavior of a twisted nematic liquid crystal layer in an electric field. *J Phys,* colloque C1, no. 3.

[VOS66]   Vos, J. J. 1966. The color stereoscopic effect. *Vision Res* 6:105–7.

[WAAC85]   Waack, F. G. 1985. *Stereo photography,* trans. L. Lorenz Huelsbergen, ed. S. Pinsky and D. Starkman. London: Stereoscopic Society.

[WALD86]   Waldern, J. D., Humrich, A., and Cochrane, L. 1986. Studying depth cues in a three-dimensional computer graphics workstation. *Int J Man Mach Stud* 24(6): 645–57.

[WALD91]   Waldsmith, John. 1991. *Stereo views: An illustrated history and price guide.* Radnor, PA: Wallace-Homestead.

[WALL53]   Wallach, H., and O'Connell, D. N. 1953. The kinetic depth effect. *J Exp Psychol* 45:207–17.

[WATA88]   Watanabe, T. 1988. A fast algorithm for color quantization using only 256 colors. *Syst Comput Jpn (USA)* 19(3):64–72.

[WAY88]   Way, T. 1988. Stereopsis in cockpit display—a part-task test. *Proceedings of the Human Factors Society 32nd Annual Meeting,* vol. 1, pp. 58–62. Santa Monica, CA: Human Factors Society.

[WEST86]   Westheimer, G. 1986. The eye as an optical instrument. Chap 4. in *Handbook of perception and human performance,* vol. 1, ed. K. Boff, L. Kaufman, and J. Thomas. New York:Wiley.

[WEST89]   Westheimer, G., and Truong, T. 1989. Target crowding in foveal and peripheral stereoacuity. *Am J Optom Physiol Opt* 65:395–99.

[WHEA38]   Wheatstone, C. 1838. Contributions to the physiology of vision—part the first. On some remarkable and hitherto unobserved phenomena of binocular vision. *Philos Trans R Soc Lond* 128:371–94.

[WICK90]   Wickens, C. D., Kramer, A., Andersen, J., Glassner, A., and Sarno, K. 1990. Focused and divided attention in stereoscopic depth. *Proc SPIE: Stereoscopic displays and applications* 1256:28–34.

[WILS91]   Wilson, H. R., Blake, R., and Halpern, D. L. 1991. Coarse spatial scales constrain the range of binocular fusion on fine scales. *J Opt Soc Am A* 8:229–36.

[WIXS88]   Wixson, S. E., and Sloane, M. E. 1988. Managing windows as transparent PAGEs in a stereoscopic display. *Proc SPIE: Three-dimensional imaging and remote sensing imaging* 902:113–16.

[WIXS89]   Wixon, S. E. 1989. Three-dimensional presentations. *Inf Disp* 5 (7,8):24–26.

[WIXS90]   Wixson, S. E. 1990. Volume visualization on a stereoscopic display. *Proc SPIE: Stereoscopic displays and applications* 1256:110–12.

[WURG89]   Wurger, S. M., and Landy, M. S. 1989. Depth interpolation with sparse disparity cues. *Perception* 18:39–54.

[YAMA86]   Yamada, H., Masuda, C., Kubo, K., Ohira, T., and Miyaji, K. 1986. A 3-D display using a laser and a moving screen. Proceedings of the 6th International Display Research Conference, Tokyo.

[YANG91]   Yang, Y., and Blake, R. 1991. Spatial frequency selectivity of human stereopsis. *Vision Res* 31:1177–90.

[YE91]   Ye, M., Bradley, A., Thibos, L. N., and Zhang, X. 1991. Interocular differences in transverse chromatic aberration determine chromostereopsis for small pupils. *Vision Res* 31:1787–96.

[YEH90]   Yeh, Y.-Y., and Silverstein, L. D. 1990. Limits of fusion and depth judgment in stereoscopic displays. *Hum Factors* 32:45–60.

[YEH92]   Yeh, Y.-Y., and Silverstein, L. D. 1992. Spatial judgments and stereoscopic presentation of perspective displays. *Hum Factors* 34:583–600.

[YZUE83]   Yzuel, M. J., Navarro, R., Campos, J., and Calvo, F. 1983. The use of commerical fine-grain films for computer-generated holograms. *Opt Laser Technol* 15(3): 138–40.

[ZENY88]   Zenyuh, J., Reising, J. M., Walchi, S, and Biers, D. 1988. A comparison of a stereographic 3-D display versus a 2-D display. *Proceedings of the Human Factors Society 32nd Annual Meeting,* vol. 1, pp. 53–57. Santa Monica, CA: Human Factors Society.

[ZUME86]   Zumer, S., and Doane, J. W. 1986. *Phys Rev A* 34(4):3373.

Index